# THE
# FARMING LADDER

*by*

# G. HENDERSON

**British Library Cataloguing-in-Publication Data**
A catalogue record for this book is available
from the British Library

# CONTENTS

# Farming

Agriculture, also called farming or husbandry, is the cultivation of animals, plants, or fungi for fibre, bio-fuel, drugs and other products used to sustain and enhance human life. Agriculture was the key development in the rise of sedentary human civilization, whereby farming of domesticated species created food surpluses that nurtured the development of civilization. It is hence, of extraordinary importance for the development of society, as we know it today. The word *agriculture* is a late Middle English adaptation of Latin *agricultūra*, from *ager*, 'field', and *cultūra*, 'cultivation' or 'growing'. The history of agriculture dates back thousands of years, and its development has been driven and defined by vastly different climates, cultures, and technologies. However all farming generally relies on techniques to expand and maintain the lands that are suitable for raising domesticated species. For plants, this usually requires some form of irrigation, although there are methods of dryland farming. Livestock are raised in a combination of grassland-based and landless systems, in an industry that covers almost one-third of the world's ice- and water-free area.

Agricultural practices such as irrigation, crop rotation, fertilizers, pesticides and the domestication of livestock were developed long ago, but have made great progress in the past century. The history of agriculture has played a major role in human history, as agricultural progress has been a crucial factor in worldwide socio-economic change. Division of labour in agricultural societies made (now) commonplace specializations, rarely seen in hunter-gatherer cultures, which allowed the growth of towns and cities, and the complex societies we call civilizations. When farmers became capable of producing food beyond the needs of their own families, others in their society were freed to devote themselves to projects other than food acquisition. Historians and anthropologists have long argued that the development of agriculture made civilization possible.

In the developed world, industrial agriculture based on large-scale monoculture has become the dominant system of modern farming, although there is growing support for sustainable agriculture, including permaculture and organic agriculture. Until the Industrial Revolution, the vast majority of the human population laboured in agriculture. Pre-industrial agriculture was typically for self-sustenance, in which farmers raised most of their crops for their own consumption, instead of cash crops for trade. A remarkable shift in agricultural practices has occurred over the past two centuries however, in response to new technologies, and the development of world markets. This also has led to technological improvements in agricultural techniques, such as the Haber-Bosch method for synthesizing ammonium nitrate which made the traditional practice of recycling nutrients with crop rotation and animal manure less important.

Modern agronomy, plant breeding, agrochemicals such as pesticides and fertilizers, and technological improvements have sharply increased yields from cultivation, but at the same time have caused widespread ecological damage and negative human health effects. Selective breeding and modern practices in animal husbandry have similarly increased the output of meat, but have raised concerns about animal welfare and the health effects of the antibiotics, growth hormones, and other chemicals commonly used in industrial meat production. Genetically Modified Organisms are an increasing component of agriculture today, although they are banned in several countries. Another controversial issue is 'water management'; an increasingly global issue fostering debate. Significant degradation of land and water resources, including the depletion of aquifers, has been observed in recent decades, and the effects of global warming on agriculture and of agriculture on global warming are still not fully understood.

The agricultural world of today is at a cross roads. Over one third of the worlds workers are employed in agriculture, second only to the services sector, but its future is uncertain. A constantly growing world population is necessitating more and more land being utilised for growth of food stuffs, but also the burgeoning mechanised methods of food cultivation and harvesting means that many farming jobs are becoming redundant. Quite how the sector will respond to these challenges remains to be seen.

'Oh, little valley, all our own,
Here is the place where beauty dwells,
 And all the joys this world has shown
Your gift of quietness excels,
 Nor would I change your stream and trees—
For jewels of the Seven Seas.'

*A GENERAL VIEW OF THE FARM*

# PREFACE

The object of writing this book is to demonstrate how a happy, secure, and useful life may be spent, on what were a few barren acres, without the toil and drudgery which are usually associated with smallholding, and that a financial return may be obtained comparable with that in any other business. The methods used, none of which are clever or original, will show that in peacetime depression or in wartime prosperity the creative work of the farmer can have its just reward, independent of tariffs and subsidies, if directed on the right lines; that great capital, special knowledge or skill are not essential, only energy, patience, and a thorough grasp of the underlying principles.

The great spate of books which have been written on agricultural reconstruction after the war, of which the author has only read the reviews, all seem dependent on some drastic action by the Government, Agricultural Committees, or some special body created for the purpose, and always at the expense of the community at large; while here I try to show that the real solution lies in the hands of the individual farmer, who only requires the time, space, and faith, which should be the birthright of all free people.

This is the plain unvarnished story of a farm and the people who have worked on it over a period of twenty years. In it will be found none of the fine writing or sentimentality about farming which is the prerogative of other writers. If it helps others to make a success of their farming it will be well justified. It is written at the suggestion of visitors to the farm, and is the outcome of an article in the *Farmer and Stockbreeder* describing the farm and the methods used, refuting the suggestion that the future of British farming lies in big farms. This article resulted in over two hundred letters requesting further particulars.

One note of warning. There is a 'farming ladder' for all who can make use of it, but remember that it is a ladder and not an escalator; it must be climbed step by step and one must be prepared to take the full weight on each rung. The two feet must be placed firmly on the good earth, or it may be a danger to yourself and to others, and although it may enable you to achieve the heights, you may never reach the top. Beware of the easy, the gaudy, the cheap, the second-hand, and the rotten. Even on our bright and shining ladder there may be times when you have to balance on one rung and reach for the next. If you have faith in yourself and the land there are only two rungs you need fear, ill health and accident. Once started you will find others on the ladder too, some above and some below: give them all a helping hand, but avoid those who expect you to drag them up and make no effort of their own. Those who would stop you climbing, or shake you off, may well be ignored.

G. HENDERSON

*Enstone, Oxford*
*September* 1943

On the occasion that the *Farming Ladder* is going to press for the fourth impression the author takes the opportunity of thanking the many farmers and others who have written to him in connection with it. That the book has achieved its purpose there can be no doubt, for a great many young people have written to say that it has given them an object and purpose for the future, for it has shown them how they may succeed in the way of life that appeals to them. The author feels that these will be worthy successors to the many old farmers who have also written to him, telling of their own struggles on the 'Farming Ladder', and against which the record given in this book has been a very humble effort indeed.

While in the past the farm has always been open for inspection by anyone who cared to make an appointment, we have had so many visitors in recent months that we have found that it is no longer possible to show individual visitors round, and we therefore now arrange a 'visitors' day' from time to time, particulars of which can be obtained by post, if a stamp is enclosed.

*July*, 1944

G. HENDERSON

# ACKNOWLEDGEMENTS

Acknowledgements are due to the *Farmer and Stockbreeder* and to the *Poultry Farmer* for permission to reproduce photographs supplied by them.

# CHAPTER ONE

## THE FARM

We saw it first twenty years ago, on a cold grey morning, lying in an isolated little valley on the eastern slopes of the Cotswolds; eighty-five acres of poor, stony land, overgrown hedges, tumbledown buildings, arable a mass of weeds, and grassland, if such it could be called, full of little bushes, or rushes in the wetter parts.

Six miles from a railway station, well off a road over a swampy piece of common land, and far from a school or village; such, probably, were the reasons why it was vacant. Yet a careful inspection showed that it had good possibilities. It was compact, lying in a ring fence, with the site of the buildings well placed. The house was small but sound. A fine spring of water bubbled out of the ground, and although sadly trodden in by pigs and cattle, it supplied every field on the farm, and there were geological indications that it had never run dry. It was a healthy soil for stock rearing, stone brash (oolitic limestone), 450 feet above sea-level, with a general slope south, and sheltered by higher hills. Most important of all, THE LAND—most long-suffering of all Man's possessions—had been badly neglected, but not robbed, as the situation prevented the selling-off of all the produce other than on the hoof. We have been told since that on more than one occasion four horses were required to drag out eight sacks of wheat through the swampy common land behind which the farm is situated.

So here was a farm that was secluded, too far from a village for stray cats, stray dogs, and STRAY PEOPLE to be a nuisance. It had a water supply free from pollution—at that time the nearest large village drew its drinking water from a stream which drained two churchyards and six farmyards, with a peculiar flavour and rich colour only appreciated by those reared upon it, who now deplore the tasteless, odourless,-and colourless product which has to be paid for by a water rate! The overgrown hedges could be cut and laid, the ditches scoured and the swamps drained. The buildings could be made sound and adequate by our own labour. A road could be built from stone quarried from the hillside (it took a thousand tons and all our spare time for ten years to make it 660 yards long and capable of taking the heavy traffic now imposed upon it). Above all, the land could be cleaned and made fertile. Everything else was dependent on this in our case, every penny of our resources would be absorbed in taking the farm, all we put on the land would first have to come out of it by the sweat of our brow.

The vendor was not available to show us round this 'Small Desirable Property' as the auctioneer's particulars and order-to-view describe it. But the lad he employed was found asleep under a hedge while his two neglected horses stood yoked to a plough. He was pleased to point out the boundaries and give any information required.

The last harvest had yielded about forty quarters of corn. Ten tons of straw and five of hay were available. The roots, such as they were, had been given to a neighbouring farmer to pen off. A few calves were reared, grazed on the common and roadside wastes, and sold as stores. A couple of sows, an odd cow, and a few ducks completed the stock. The last tenant had been there for ten years, was reputed to have come with nothing and gone away with £1,000, a story we did not doubt in view of the unparalleled prosperity agriculture had enjoyed during the Great War and two years after it, which left the farms and farmers poor in everything except money. It was very certain that on this farm far more had been taken out than had been put in.

Leaving word with the boy that we would come again three days later, advising the time by telegram so that the owner could be on the spot, we went on to the village inn a few miles away, not with the sole intention of imbibing strong drink, or spending money on a lunch (although we did both), but so as to acquire local knowledge. The parson or squire might be able to tell a little if so disposed, but no secret is safe from the village gossip, and elderly labourers, having nothing to look forward to, forget nothing of the past, and will repeat themselves over and over again when the beer is flowing.

A few pints produced a complete summary of the habits, morals, ability, and financial standing of every occupier and landlord back to 1837. All, it appeared, had done well; one farmer born there in the 'fifties had had twenty-two farms in his time and left over £40,000.

The name Oathill Farm, for which we could see no simple explanation on the site, was invariably pronounced 'Olt-hill'. This puzzled us at the time, but pondering later we recognized the truth of the saying, 'Words are the only things which last for ever'. The Brythonic 'Allt', still used in Wales and the west of Scotland, meaning a little stream in a valley, correctly described the farm. Some solicitor's clerk drawing up the deeds perhaps a hundred and fifty years ago might spell it 'Oathill' as a familiar word nearest to the pronunciation as he heard it, but the local labourers, who have never heard of philology, still use Allthill, as their fathers did before them. The meaning has been lost since Roman times, but waits for the understanding of those who have ears to hear.

If any reader would care to check this, look at any Oathill, Oatley Hill, or Holthill, on the Ordnance map, and he will find that in every case there are a hill, a little valley, and a stream.

The farm had originally been divided into four large fields, from the open field system of about 1800, for the familiar four-course rotation, of winter corn, roots, spring corn, and clover. One of these fields had been subdivided in 1890 to provide a permanent grass field, and this general layout can still be seen on the map. But no rotation seemed to have been followed for at least ten years. The method, if such it could be called, was to leave the land in grass, usually a temporary mixture of rye grass and clover, from which the latter soon disappeared, and then plough a strip for winter corn, and then, when too late for autumn wheat drilling, start ploughing again for spring corn on another strip. After two or three corn crops an odd patch would be put in roots and the rest put down to grass again, complete with the arable weeds which had been encouraged by the corn crops. Much of the land in this district is still being farmed on this 'neolithic' rotation, but without the saving grace of being left in grass from time to time, growing poor crops of corn, corn, corn, then potatoes to rob the land a little more, followed by more corn; a procedure which in the long run impoverishes the farmer as well as the land.

At Oathill as many as six crops appeared as odd patches in one field. All the arable was a mass of couch grass and docks. The permanent grass, mostly bent and broom grass, had tumbled down; there were no signs of clover, and little thorn-bushes were springing up everywhere. The stream, which flowed so clear in places, flooded about three acres; these were covered with rushes for want of a little cleaning, excellent perhaps for the ducks, but a wicked waste of good land, a breeding-place for fluke and other parasites, and an eyesore to tidy farmers.

The buildings consisted of a large stone barn, roofed with Cotswold slates, which is simply the oolitic limestone, split by frost and shaped by hand; a stable, with loft over, for four horses; an open cattle shed; two small loose boxes; three pigsties; a two-bay cart shed. The first two were in a good state of repair, but dark, and with bad floors. The rest were in the last stages of dilapidation, with leaky thatch running water on the walls and washing out the mortar, and all threatening to fall down at any moment. What is worse for stock than a damp, dark, building? The accumulation of the manure in the yard was so deep that the anaemic-looking cow upon it was able to eat the thatch, or the corn sprouting from it. Not a gate on the farm would swing, several were broken beyond repair, and others missing altogether.

However, to return to the inn. The last round of drinks, by which time we had nearly bought the pub up, produced the information that the farm was also haunted. The occupier was lodging in the village, as he did not care to sleep there at night on his own, and one housekeeper had left the same day as she arrived. Believing, in those days, in very little which could not be proved by chemical analysis, this only interested us from the point of view of driving a better bargain. In passing, it may be mentioned that the only manifestation we have seen which could be mistaken for a phantasm takes the form of a large white owl with only one claw. Living in the barn, very tame, and largely dependent on vermin we trapped and left for him, he has been handled when gorged with food, and he purred like a cat. Living for something over twenty years, and willing to clear up four or five rats at the time, this kindly spirit of the wild illustrates what a service his feathered brethren must render to agriculture. He finally fell a prey to a poisoned rat baited by the official ratcatchers of the local War Agriculture Committee.

Tramps, we were also told, were a great nuisance, lying about all over the place, and the local people did not care to go down there after dark. Gipsies were reputed to camp on the common and steal everything they could lay their hands on. A pleasant outlook for a lonely farm!

Fortunately, the writer had learned how to deal with all and every kind of vermin which infest the countryside, so we did not worry overmuch. Normally tramps only camp on farms which are not occupied, and are afraid of dogs and men. They depend largely on begging in the villages when the men are away at work and the women are afraid of them, and give them money and food which they can ill afford. There are only two things a tramp should ever be offered: work, if you are very kindhearted, an offer which he does not want; or personal violence. If you must leave your wife alone in the house, buy a second-hand policeman's helmet and hang it in the hall, and leave the door open. Gipsies are different. Play on their superstitious fears by hiding their death sign where you know they will find it, and they move on without taking their horses out. Destroying all the sallow willow is also a good method of discouraging them, because it is from this that they make the clothes-pegs which are hawked round the villages.

Having advised the time of our arrival, as arranged, we duly arrived at the farm again to find the familiar buff-coloured envelope stuck under the door-knocker, showing that the owner had not been at home to receive it. This simple gesture, we understood, was the old English farming custom of pretending that one is not too anxious to do business. Just as a third-rate solicitor, bank manager, or civil servant is always busy writing when you are shown into his room, while the really big men in business or the professions are always ready to meet you at the time appointed and deal immediately with the matter in hand.

However, as we had come to inspect the house, and had no time to waste, we intended to get on with it. A ladder from the buildings, a loose window catch on the second story and a strong clasp knife helped us to make an entry. All the lower windows were shuttered and barred, and the front door was locked, double bolted, barred, and chained. These precautions amused us so much that we could not resist the temptation to place the telegram in the centre of the dining-room table, before quitting the house and carefully shooting the window catch, and before returning the ladder to its proper place. On his return the owner would have to think the pet ghost had been busy again.

This visit confirmed our previous impression that the house was small, but sound, with four bedrooms, two sitting-rooms, a kitchen, dairy, and outhouse. In the outhouse were a good pump and sink, together with bread oven and copper. We could add a bedroom and bathroom without any great expense, when time and money permitted. The usual medieval convenience at the end of the garden could also be replaced by indoor sanitation as there was plenty of water available.

The garden contained an excellent crop of nettles, tin cans, and ashes, with a few mildewed gooseberry bushes and plum suckers. The lawn, over which the pigs roamed, and where the ducks were penned, was redeemed by two fine spruce firs, the relics, perhaps, of some bygone Christmas when some happy family celebrated the festival here in the time-honoured way.

An orchard adjoining the house contained a score of old apple-trees, some already tumbled down, others in the last stages of decay, and only one worth keeping.

The rickyard and rick sites elsewhere covered a couple of acres, with old rotten straw, dumps of thatch, old implements, thistles, and other rubbish. This is still far too common on many farms, indicating that the farmers have more land than they deserve; they ought to be able to put it to better use than as a site for the rotting down of good straw, which should have been long since returned to the land; it is a pity if the rickyard cannot grow something better than weeds.

Another walk round the farm, summing up its faults and failings (which I have given in detail) convinced us that we could make a success of it. It only remained to find the elusive owner and complete the deal.

The assurance and confidence of youth! The elder partner was under twenty years of age, and the younger not quite eighteen. Capital £150. Knowledge and experience? Four years' general farming as a pupil for one, two years' engineering and business for the other. Our greatest asset was a mother with absolute faith in all her children, believing them to have inherited in some small measure the ability of their father, who went to London as the penniless son of a small Scottish farmer to build step by step a successful business. This almost failed at his early death, and she was left to rear and educate five children under sixteen. All were given the best education she could afford, a free choice of trade or profession, but in which they would have to make their own way on leaving school. The older sons safely launched, the two youngest, choosing farming, were backed to the limit of her resources, with a farm bought on mortgage, a little working capital borrowed from an old friend, and a supreme trust in Providence. What more could anyone want, excepting perhaps a mother who was also prepared to pioneer in the wilderness and keep house for her sons. In full measure we had all these things.

It should perhaps be mentioned that any so-called expert on farming, whether farmer, auctioneer, land agent, or county organizer, would have predicted utter failure for this venture. Prices were tumbling, old and capable men were said to be losing money. It is quite certain that had there been a War Agricultural Committee functioning in those days we would never have been allowed to take the farm at all. A bank manager poured scorn on all my plans and schemes, refusing to grant even a £50 loan to tide over a short period. I blushed as I stood before his desk and he talked to me like a delinquent schoolboy, yet in more recent years I have lounged in a chair while a bank director listened with careful attention to my views on financial stability in farming, which have not changed in the smallest detail over the years.

*PERMANENT PASTURE, 1924*
*THE QUARRY, SHOWING THE FOUR INCHES OF SOIL ON*
*SOLID ROCK*

*THE GEESE COME HOME*

*A BREEDING FLOCK*

Others may be tempted to do likewise under similar circumstances, so ignore the gloomy predictions of those who have looked on the grapes which are sour. Given the will there is hardly anything which cannot be achieved. Remember that the aerodynamic experts can demonstrate, with all the resources at their disposal, that nothing the size, shape, and weight, and fitted with such inadequate wings, as a bumblebee can possibly fly. The bumblebee, not knowing this, but having the will, does so quite comfortably, every day, collects a little honey, and at the same time performs a service to agriculture which cannot be measured in millions of pounds, by the fertilization of red clover which is the basis of rotational farming.

From what did such confidence spring? From his earliest youth, the writer felt that there was only one thing worth doing on earth—farm it! I had listened for hours on end to my father's tales of his boyhood days in Scotland, visualizing the windswept mountainside where for eight generations our family fought against the rocks and heather encroaching on the hard-won acres from which a living could only be scratched by incessant toil. An old lady who lived with us used to tell how in the Crimean War her father earned seven shillings a week, with bread at a shilling a loaf, as a labourer in Essex. All this could leave me few illusions as to an earthly paradise on the land, compared with the comfort and security of the home in which we were born. Yet I never had the slightest doubt, and my younger brother shared my enthusiasm and assurance that we would some day become farmers.

All that remained was to find the ways and means. Reared as we were under the smoke pall of a great city, with only a few yards on which to keep poultry and rabbits, and our knowledge of the country gained only at rare intervals, it would seem that we had a long way to travel to realize our ambitions. However, as we got older we spent many happy days in the country, and helped on a farm in the last years of the Great War.

In my last year at school I took the full course from the Agricultural Correspondence College, which was then at Ripon, Yorks, and is now at Bath. It was well worth the ten guineas it cost, running to some hundred thousand words, a masterpiece of condensation of fact, and without a single line which did not teach me something. I should perhaps mention that I have no interest whatever in advertising this particular college; others may run a similar course, but I am convinced that in this class of training one can only take out as much as is put in by study and application. Personally I almost committed it to heart.

On leaving school, my eldest brother asked me what I intended to do for a living, and told me that my foolish dreams of becoming a farmer should be given up. Seeing that no capital would be available, I should be condemning myself to a life of unremitting toil, in mud, snow, or dust; no holidays, no security, poverty and want, saving all the years for a set of false teeth and a coffin, or finishing in the workhouse. On the other hand, as I had done fairly well at school, for a premium of £50 a sure position could be obtained in the more gentlemanly occupation of a clerk, with security, steady promotion, and a pension at the end of it.

A black moment. Were all my dreams to be dashed to the ground? Toil and hardship meant little to the adolescent finding his strength for the first time, but I did want to make a home for my mother in the country within a very few years. No! I would become a farmer by work, study, and prayer, or die in the attempt. 'Let me have the £50,' I said, 'and I will be a farmer before I am twenty-one.'

I then wrote to the farmer for whom we had worked in holidays, and arranged to work for him in return for experience. But I soon saw that this was not the place to learn what I required. A kindly, decent old man was this farmer, but always behind, or in a muddle with the work. No rotation of crops was followed, no balanced rations fed, or any of the other things I had expected to find from my correspondence course in agriculture. I learned later that very few farmers in every thousand at that time used the resources of science which were at their disposal. Much of the labour employed was useless, and from such farms nothing could be learned. The animals were always sick or ailing, the crops spoiled by the weather. This farmer, moreover, never grumbled or found fault with my work; after ten years at school I expected criticism and would have valued it. He told me I was wasting my life in agriculture, advising me to go back to the town or, since I loved the country so much, to win a scholarship, go to college, and become a County Education official—he had been told they earned as much as five pounds a week, without soiling their hands—more than he had ever earned in his life, and therefore an incredible sum to one overburdened with debts and mortgage interest, and with seldom a shilling he could call his own.

I thanked him for his advice, but I had no objection to soiling my hands and intended to earn a great deal more than the sum that seemed so much to him, as capital and a wage had got to be found. But first, by hook or by crook, I had got to learn my trade.

I advertised and answered advertisements, spent a few pounds on travelling to interviews, and learned just a little by talking to farmers and walking round their farms.

Paradoxically I found it very difficult to enter this bankrupt industry (bankrupt, that is, in the estimation of the farmers' leaders). Apparently farmers required from £50 to £300 premium to teach one how to lose money. At one end of the scale they looked for cheap labour that would pay for the privilege of working, and those at the other end of the scale spoke highly of the hunting, shooting, and fishing in their district, and hardly mentioned their farming.

At last the right place was found, with a young progressive farmer, who had modern methods of rationing, recording, manuring, and cropping, and good pedigree stock: no rich man's hobby, his farm, but a sound business concern. He was asking a premium of £60, while by that time my financial resources had dwindled to £40. Putting all my cards on the table, I talked him into taking me. It appeared he was shortly moving to another farm, a long way from the railway station, to which the milk would have to be taken every evening. None of his men would take it, as they would not get home till nine o'clock, for they lived in the opposite direction. This was too late, seeing that they had to start again at five in the morning. It was agreed that I should drive the milk after the men had gone home, and in return, if satisfactory, my premium would be returned as wages after six months.

On that place I learned how to FARM, which means a great deal more than the words convey. Good farming is the cumulative effect of making the best possible use of land, labour, and capital. This must not be confused with neat and tidy farming only, which is often uneconomic, although, of course, efficient farming will always have an order and purpose which is apparent to the discerning eye.

Everything was of the very best of its kind: pedigree horses, both light and heavy, cattle, sheep, pigs, and poultry. I learned that one really first-class beast leaves a bigger profit than a dozen average animals or a score of indifferent cattle. All consume the same quantity of food and require the same labour, but the margin between cost and selling price bears no comparison. How well that knowledge has served me!

On that farm too some of the finest judges of their day were frequent visitors, and I owe much to their kindness in teaching me how to recognize the best and not be deceived merely by show condition. What a contrast between the able and talented man, who so freely shares his knowledge, and the ignorant labourer, so common in farming, who is afraid that someone might learn from him and profit by it.

It was part of my work to keep the books, and from them I learned the fundamental reason why such small profits are made in farming. True the farmer cleared about £1,000 a year, but there was £60 an acre invested in land, stock, tenant-right, equipment, and working capital. The total capital was turned over once in three years, in which period with ordinary business it might be turned over thirty times. Therefore the solution must be intensity of production. The farm which could turn its capital over once a year with the same margin of profit would pay three times as well. British agriculture with invested capital of some £30 per acre and an output of £7, turning its capital over once in four years, was not, and could not be, a business proposition for the average farmer. So simple! I could think it out for myself at the age of seventeen, yet I suppose ninety-nine farmers out of a hundred would rather farm a thousand acres extensively and probably lose money than a hundred acres intensively and make some. True, money in farming is not everything, but we have never learned to farm without it, as land should be farmed and maintained, with fences, roads, buildings, and the wherewithal to reward good and loyal service on the basis it deserves. Carried a stage further, we have since learned that, providing the output per acre is really high, it does not matter how extravagantly you farm, you will still show a profit. On the other hand, however carefully you farm, if the output per acre is low, you will make a loss.

On the practical side I was given an opportunity to try every job on the farm, and learned to milk, drive a team, stack and thatch, and prepare animals for show and sale. In some respects I felt that I was not getting all the practical experience I would have liked, because the busier we were haymaking, harvesting, or the like, the less I seemed to have to do with it. The Boss, knowing that he must be right on the spot when important work was in hand, tended to send me to deal with little matters of business he would normally have attended to himself. I have realized since that in deputizing for him in showing prospective customers stock he had for sale, or in driving a bargain for him, I was acquiring knowledge which would some day be invaluable to me.

Working from five in the morning to nine o'clock at night, I had few opportunities for spending money. For this I was truly thankful, as my personal expenditure could be reduced to fourpence a month on hair-cutting and as little as possible on essential clothes.

I was very happy in my work, finding no toil or drudgery in it. My employer sometimes took me to shows and sales. He also taught me to ride. He was a very clever horse-master, but weighing fourteen stone and standing six feet three inches, could not ride the lightweight ponies and horses for which at that time there was a great demand. With only a few old crocks, mares, and foals, left over by the army buyers during the Great War, anything that could be hunted sold for a high price. Weighing in those days little over eight stone, and taking very easily to riding, under such a capable master, I think I proved quite useful to him. The wiles and guile of such a man in selling a horse were an education in themselves. He would have one broken-in in a few summer evenings.

Then when cub-hunting started and the meet was close to the farm, he would say, 'George, we will go to the meet to-morrow'. Off we would go, perhaps before five o'clock; the only justification I can see for fox-hunting is the ride in the early morning. Arriving in good time, he would look for a fairly easy jump, and telling me to follow him, would put his magnificent heavy-weight hunter at it, and mine would follow in the excitement of the moment. Doing this several times, until he was sure the young horse would jump without question, he would then say, 'Hang about here, after they have finished, until I come'.

In due course he would meet someone looking for a horse. 'Ah, yes,' he would say, 'I've got just what you want. I wonder where my pupil is?' Knowing full well where I was he had little difficulty in finding me. The horse having been duly inspected, the customer would ask if it could jump. 'Oh yes. Jump anything. Now what would be suitable?' Here he would look all round the country, but never at the jump the horse had been over several times. As often as not the customer would suggest the obvious place; and I would be told to put the horse over it, doing so as nonchalantly as possible. Thus another horse was sold. He had a hundred similar tricks, but I never knew him fake a horse, or sell one as sound if it was not. He considered that a perfectly trained hunter was wasted on ninety-nine hunting people out of a hundred, but for the hundredth man he could always find a beauty. His own was a miracle of co-ordination between human brain and equine muscles. Sometimes he would let me ride it home from the meet or at exercise, and I found it a unique experience. With a mouth like oiled-silk, he would start off at any pace on a named foot. So beautifully balanced was he that he could correct the rider's weight in the saddle, so responsive that a rabbit hole could be sidestepped by a twist of the wrist; he could change his feet at the touch of the leg. Approaching a jump his stride could be shortened or lengthened so that he always jumped at a distance equal to the height of the fence, clearing it with effortless ease, to the comfort and security of the rider. Such a horse could be sold for five hundred guineas, and ruined in a few weeks by some wartime profiteer taking up hunting for the first time.

But these interludes were few and far between, and I always felt a little guiltily after a day's hunting that I was neglecting the serious business of learning my trade. The financial side of hunting made little appeal, as I saw many fooling away the easy money on wine, women, and racing. It also enabled me to form first-hand opinions on a sport which has survived from the dark ages and must be considered an anachronism where modern farming is concerned.

Ethically it is wrong to inflict unnecessary suffering for the gratification of a mere thrill, and for that reason fox-hunting is a disgrace to the civilization of a country that permits it, and a reflection on the mentality of the people who take part in it. To contend that it provides employment is to put it on a level with crime and lunacy, for which the same thing could be said. The work and trade associated with the sport could be put to more productive and creative use.

Riding can be enjoyed without hunting. It is to this day my greatest pleasure, and it must be remembered that only about five per cent ride straight to hounds, the rest of the field career round the lanes and through the gates. For the good horseman there is always the drag-hunt, which almost invariably develops into a steeplechase. (This, however, makes it unpopular with those who only require a little social activity, the admiration of the crowd, the pomp and pageantry associated with hunting.)

So the days, weeks, and months slipped by. I learned something every day. The men on the farm were very helpful. I never hesitated to flatter or bribe to obtain their knowledge and skill. When one old man complained that the

Boss ought to send him out a pint of beer each day if he had got to teach me how to thatch, I instantly offered to buy him a barrel on the day I was satisfied I could thatch as well as my instructor. Never was I so thoroughly taught. Others hearing about this went out of their way to impart their knowledge and they found me a ready pupil. Strange as it may seem there are many in the industry who will not help. Quite recently we offered a rural craftsman, specially skilled in his trade, £20 a week to teach some boys, and although there was no question of him fearing competition, he refusal on the grounds that it had taken him sixty years to learn and he would rather his trade died with him than that others should acquire it easily.

That I would offer to wash the dairy utensils for the dairymaid so that she could keep a date, or stop up all night to help the poultry-maid to pluck and truss poultry, was attributed to my kindness of heart. Little did they know it was simply determination to master every task on the farm, whether outside my interests or not.

A few weeks before my year was up, a new pupil came from the Harper Adams Agricultural College. Having had four years there he was now ready to gain practical experience. In my opinion he was doing things the wrong way round, but nevertheless, from him I learned quite a lot of the theory and scientific aspect of farming, and also the standard text-books with which one should be familiar.

Altogether this was a very happy and profitable year. I finished with £35 in hand, and a very kind and generous reference, in which I was described as 'the gentlest and most efficient milker' the Boss had ever had, and 'a first-class horseman with either heavy or light animals'. In view of the unfailing kindness and patience he had taken to teach me, I am sorry that I have been unable to get in touch with him again and obtain his permission to mention his name.

Once more I was up against the problem of finding a suitable situation; the faster British agriculture slipped down the precipice of postwar depression, the higher the premiums the farmers asked. My experience counted for nothing. One farmer told me he had lost £7,000 in the last year in depreciation alone, yet apparently he felt qualified to teach, or did he really want the premium of £200? This surely would only be a drop in the bucket, as he gave no indication of changing his methods to meet the times, but could only curse the Government for repealing the Corn Production Act, under which prices were to be linked with wages.

Then I saw an advertisement requiring a young man to milk twelve cows on a small farm in Essex. I applied, and obtained the situation subject to a month on trial.

It was a small grass farm, heavily stocked with poultry and pigs, and with the cows to keep the grass down. The farmer had little interest in his cattle, apart from the milk cheque, but was a first-class poultryman and did quite well with the pigs.

By taking this job I felt it was a descent to the ranks of the farm labourers, and looked round for ways and means to remedy the situation. Carefully laying my plans, I put my whole heart and soul into the job, looking after the cows as they had never been looked after before.' At the end of the month on trial, the farmer said he was satisfied with my work and hoped I liked the situation. I told him at once that I did not intend to stop. At this he was very disappointed and offered me an extra shilling a week, raising me to six shillings together with my board and lodging. This also I refused, and was then asked what I wanted.

My conditions were that instead of milking at 7 a.m. and 3 p.m., it should be done at 5 a.m. and 5 p.m., which would give me four hours free during the day to work on the pig and poultry section and learn all about it. I also required seven and six a week. This extra half-crown I would justify by producing an extra two and sixpenceworth more milk *daily* if permission was given to use the feeding-stuffs in hand by feeding according to yield, and not the same all round, as was his method, giving too much for the dry cows and low yielders, not enough for the deep milkers. Also I required one whole day off per month, not weekly half-holidays between the milkings on Saturdays as was the local custom. The object of this was to enable me to interview other farmers when the time came to try for another place.

Telling me that I would soon get tired of my proposition, he agreed to try it. And so while doing my job as a dairyman I made the opportunity to learn how to run a successful poultry farm. Poultry in my experience requires as much knowledge, care, and attention as any other class of stock, yet more people go into this branch of the industry than into any other, knowing nothing whatever of even the guiding principles on which a poultry farm should be run.

An incident which clearly illustrates this point occurred a few weeks after I went to this farm. A gentleman came with his poultry manager (note the manager) to buy some breeding stock. In taking them to see the birds, we passed a pen of large Buff Rock capons. 'What are those?' inquired the gentleman. 'Capons, sir,' I replied. Nothing more was said until they had inspected the breeding stock and selected what they required. And then, when they were about to leave, the manager said to his employer, 'What about having a breeding pen of those capons?'

A capon is, of course, an emasculated cockerel, and useful only for fattening or rearing chickens. Can anyone imagine the manager of an ordinary farm not knowing what a bullock was?

Telling this story to my employer, he said, 'That's nothing, I once sold a man a hen and setting of eggs. Meeting him later and inquiring how they had done, I got the reply, "She hatched ten, but I supposed she wouldn't have enough milk for all those, so I drowned four out of the way".'

Every possible source of revenue was exploited on that farm. We had a dozen breeds of poultry, four of ducks, two of geese, and turkeys. Even peacocks added their shrill cry to that of guineafowl and the challenge of golden pheasant. Hatching eggs, day-old chicks, broodies, fattened cockerels, and breeding pens brought in a steady income, the rare birds being as profitable as the domestic poultry. Peacocks would sell at £5, yet took no more food than a cockerel, reared under broody hens, followed by capons, as they require heat for six months, and the hen gets tired of them after a few weeks. With the secretive attitude of the countryman my employer would never admit he cleared a profit. Yet I calculated he cleared £400 a year on the poultry, against £100 on the cattle, and as much on the pigs. All this on twenty-seven acres, while the great arable farms of the district were falling down to scrub, as their owners went bankrupt. I saw here quite clearly that it is not the acreage you farm, but the intensity of production you maintain, which determines the financial success of the venture.

As week by week went by I studied the advertisement columns in the agricultural papers, always searching for a suitable vacancy that would enable me to take another step on the uphill path to becoming a farmer in five years.

At last I found it. An elderly sheep farmer in the far north required an assistant. I applied and enclosed a stamp. The reply came, that although he liked my letter, I was too young, and a year's experience of lowland sheep was insufficient. I wrote again saying that a young horse is easier to break than an old one, that sheep were in my blood, my ancestors having been reared on ewes' milk for generations in Scotland, and begged for an interview.

My request was granted, and I travelled through the night for the interview. How the great range of cloud-topped mountain stirred my blood, the purple heather and white foaming streams, the scent of moss and cool sweet air. What a contrast from the flat desolation of the Essex marshland where I had been working!

No-one was at the station to meet me. So, buying an Ordnance Survey map of the district, I walked the six miles over the hills to the farm. The more I saw of the country the more I loved it and felt I simply must get the job. Here it seemed to my boyish imagination was a real man's job, shepherding on these hills.

My reception at the farm did not seem too cordial. The farmer eyed me up and down, like a horse he might be buying. The conversation on his part was restricted to 'Aye' and 'So', and an occasional 'No', while his wife sat silent without saying a word. However, I said my piece, and hoped for the best.

Then, after a ten-minute silence by the clock, he said, 'We'll look round'. Outside, he silently showed me round the steading, leaving me to make what I trusted were suitable comments. At last we arrived at the Dutch barn, and stood under it, as rain was now falling in a steady downpour. Looking across a field he suddenly said, 'What breed of sheep is that?' I could only just see two sheep through the driving rain, and on the spur of the moment, and for no reason at all, except that I had heard they were kept in that district, I said, 'A Swaledale'. A breed I had never seen and could not have described.

Not a muscle of the farmer's face moved. Then he said, 'Is yon a lamb or a ewe?' Once again I guessed, and said, 'A lamb'. Turning to me, with a smile, he said, 'You'll suit me weel, I shall no bother to take up any references'.

Taking me back into the farmhouse, we sat down to a substantial meal, which his wife had prepared, and over it I learned that the farm extended to some two thousand acres, fifty arable, a hundred enclosed grass, and the rest open moor.

The stock consisted of 650 Blackfaced mountain ewes, and thirty pedigree Swaledales for ram breeding. These were his pride and joy, and I realized what a lucky guess I had made in naming the breed. According to the law of averages it might well have been a Blackface when I said a Swaledale, as I did not then know the difference. Apart from the sheep there were a few cows which reared their own calves, half a dozen working horses for the arable, and the same number of rough ponies for riding round the moor.

Returning in the train that night I took myself to task very seriously. 'George, my lad, there is not going to be any more jobs got by guessing. You will now study the photographs in *The Farmer and Stockbreeder* year books, until you can recognize any breed at sight.' So to this day I can distinguish any of the thirty-six breeds kept in this country, and nearly all the crosses between them.

By the next morning I was back in Essex, and gave fair notice to my employer. He called me all the names he could lay his tongue to, which hurt me very much at the time, but on reflection I realized that a man who is very annoyed in those circumstances has really valued one's services very highly. We parted, certainly without any animosity on my part, as I had had the opportunity to show I could manage a herd of cows without supervision, had gained a good all-round knowledge of poultry and bee-keeping, and learned to do a number of useful odd jobs, such as concreting, gardening, and hedge-laying, on which that man was very efficient.

In due course I arrived again at the sheep farm, and hid my lack of knowledge as far as possible by outdoing everyone on the farm in taciturnity, never using two words if one would do, not that unless really necessary, and that only after due reflection. This I found invaluable as a form of discipline, as I usually talk too much, and it gave me far more time to think, and forced other people into loquacity for sheer self-defence.

I soon found that my new employer was a very sick man, and drinking hard. Right from the start I had a lot of responsibility thrust upon me. I had to carry all the orders to the men, and report on the work which had been done. On Mondays the farmer went to market and returned drunk. On Tuesdays he was too ill to get up. On Wednesdays he was in a terrible temper. On Thursdays silent. On Fridays apologized for anything he had said on Wednesday. For the rest of the week was one of the nicest men I have ever known.

All this I found very wearing; but after a bit I learned to find out by Sunday night all he wanted done in the following week. On that evening I would often sit up till midnight, absorbing the accumulated knowledge of his fifty years' experience, and then going to bed with a heavy heart, knowing that for the next four days everyone's life would be made a misery, by a man bolstering up his failing health with spirits. For a drunkard is always a very sick man.

Under my new system I gave the orders to the men throughout the week regardless of what he ordered or countermanded. I often made mistakes through lack of knowledge or sudden changes of weather, which I had not learned to forecast. But on the whole we managed better, and were able to keep labour, which had not been possible before.

Once away from the house, what a grand life it was. With a whole day to work and the horizon for the farm boundary. We sometimes went weeks without seeing a stranger. The post and essential supplies were left at the shepherd's cottage a mile away. There was of course a full day's work to be done on the enclosed land as in general farming, but one had only to glance up, to see the great flock grazing on the green surrounding mountains, which shut in the farm like the rim of a saucer. This never failed to thrill me: some age-old instinct, I suppose, from generations of men whose lives must have been spent in the care and protection of the flocks and herds from neolithic times.

So deep does that instinct lie that the best type of shepherd and flockmaster develop almost psychic powers in the care and protection of their charges. Once on a still and sultry night my employer woke me at one o'clock, and told me I must move the sheep from the lower ground. I was out in a few minutes, but could not detect any reason for his decision. A storm might be threatening, but as yet there was no rain. However, orders were given to be obeyed. Calling the dogs I set off; and got my first surprise at the river running through the valley. It was brimfull and beginning to sweep over the wooden footbridge.

Setting the dogs to their task I no sooner had the flock moving than I could hear the river roaring in spate, yet still not a drop of rain had fallen. Then as we mounted higher, continuous lightning began to flicker and flash in the sky on the other side of the range, and I realized that a great storm was raging on the other side and the whole watershed was feeding the torrent which now overflowed the banks.

Then with my sheep safe, I became anxious; what of the shepherd below? I had cleared my section, the flock below would be in greater danger. Running hard for half a mile, being above the rocks and on heather, it was a great relief to find the next flock coming steadily up under the masterly control of the shepherd's dogs. But where was the man himself? With the continuous lightning the whole landscape was illuminated, yet not a sign or a signal, although the dogs co-operated under perfect control, compared with the ragged work mine had put in at an unaccustomed task. Normally they were used to bringing sheep down the hillside to be penned, and never up, as when turned loose the flock find their own way to the higher ground. Then the storm broke, and dashing for shelter under the rocks, for a thunderstorm is no joke at fourteen hundred feet, I thought no more about the shepherd, a man well able to look after himself under any circumstances when his sheep were safe.

At dawn the storm was spent and down through the rising mist which shrouded the mountainside, I slowly made my way. Reaching the river, I found it still a hundred yards wide, although subsiding, and the bridge swept away. On the other side stood the shepherd.

The moment he saw me came the hail, 'Have you seen my sheep?'

This surprised me. 'They are where you put them.'

'Are they safe?' he shouted again.

'Yes,' I replied, 'you know very well.'

'Thank God! Where was my faith? I could no' cross, the bridge was awa'. I prayed for yon to be guided.' His two dogs, hearing his voice, came trotting along the bank on my side of the river looking for a place to cross. I was left to wonder at the prescience which warned my employer of the approaching danger which woke him after midnight; and the telepathic communication between the collies and the shepherd in response to the intensity of thought and concentration called prayer.

Then one day the master had a stroke and collapsed completely, and for seven whole weeks I had complete control of the farm. It so happened that with the exception of the shepherd we had all new labour at the time, and so they accepted my authority without question, although I was only eighteen at the time.

How we worked during those weeks, and I learned to trust and rely on the best type of labour, who toiled unselfishly without thought of reward and at considerable personal risk simply to do their job.

Snow is the greatest danger, and at one time all worked for fifty hours straight off, driving the flock in a raging blizzard, at 1,400 feet above sea-level, with the snow in their faces, to save them from the great drifts which trapped many hundreds of sheep on other farms in the district.

All day long, one Sunday, as we rested before the fire, or looked out of the window, the weather steadily deteriorated, until by seven o'clock it was obvious that the sheep would have to be driven from the deep corries in which they would shelter from the rising wind and falling snow.

The shepherd, two men, and myself, set out, heavily clad and with scarves tied over our mouths as the horizontally driven snow made it impossible to breathe without them. Each with the long northern crook, or cromac as it is called, made from seven foot of ash, and two dogs at heel, we toiled up the mountainside.

On the top the gale raged at seventy to eighty miles an hour, but in deathly silence as the great plume of snow muffled every sound. So strong it came that we could not stand against it, but crawled over the ridge, the faithful collies keeping close to our sides. Then we went diagonally down till each found the deep, wide, riven gully into which the sheep were already gathering for shelter. These were hustled out, and would move along to the next in the hope of finding shelter, so on each of us depended the whole safety of the flock. As no man could move along except with the greatest difficulty, each would have to guard his own, or have the whole flock smothered under perhaps twenty feet of snow. Sheep can of course live for some days under snow; the record is about six weeks, but they catch cold when released and develop what is called snow fever.

Although on the farm I was now accepted as the foreman, having passed all the orders from the farmer to the men, and now he was ill, making my own decisions, on the open moor when there was work to be done, it was the shepherd who took charge.

There were four open corries to be guarded, so leaving me at the first, he led the other two men on to their places and finally battled to the most difficult of all, where the full blast of the wind drove the snow in great drifts which would remain all winter.

Rough, roofless stone shelters, rather like grouse butts, had been set up at some time. These had now filled with snow, but by scraping it out it was possible to get a few minutes' respite from the wind, which in spite of the cold, or because of it, felt like the breath from a blast furnace and driven sand the moment you faced it.

The collies too would bury themselves in the snow, and had to be routed out every time the sheep came silently along like bundles of cotton-wool as their fleeces trailed in the snow, travelling before the wind on the open mountainside which was swept free, but getting deeper and deeper in the places where they sought shelter. A Blackface mountain sheep is almost helpless in even a two-foot drift and has to be dragged out by brute force, or have a track trodden for it to enable it to get out.

And so in the bitter cold and darkness of a long winter's night we exercised unceasing vigilance until dawn. Then the shepherd moved back with the wind to his first man, who went on to take the shepherd's place. And then to the next, changing over again, and finally to me. In this way one could go back to the farm for a meal, report all was well on the hill, and return with food and drink for the others.

At midday the weather moderated; but in the late afternoon renewed its onslaught with unabating fury, and the wind went back from east to north, always a bad sign. So we were then faced with another night on the mountainside, as the sheep started to come again to the hill. They had grazed on whin and gorse all day on the lower slopes, which were only partially covered with snow, the drifts forming only in the deep valleys in the bottom, and the sheltering places on the top. For a mountain sheep farm, or hirsel, is selected in such a manner that food and shelter can be obtained by the stock, but no protection can be devised from snowdrifts, as sheep seem to have no self-preserving instinct in regard to them; which is strange in such intelligent and self-reliant creatures.

On the Tuesday morning the flock was still safe, but the shepherd was suffering from exposure and exhaustion, and it was agreed that the other two men should take him back to the farm, the old man first exacting a promise from me to continue minding the sheep until dark, when I would be relieved if the weather did not improve.

With the resilience of youth, fortified by rum and milk, from which the greatest benefit is derived by one who normally never touches alcohol—proving it can be a good friend, if sometimes a bad master—I was able to continue slowly moving backwards and forwards on the mile-long stretch; though with a falling wind, going round with the sun, the danger to the flock became less every hour. At dark they were safe, and I slowly picked my way home, very, very tired but truly thankful that I was equal to the hardest physical task that might come the way of any farmer, and compared with which the hardest day's threshing is but child's play, because it does not go on all night. Reaching the house at nine o'clock, exactly fifty hours from when I left it, I went to sleep immediately I sat down in the warm kitchen.

The next day we rested, all ordinary work being at a standstill, but then we had to be hard at work again helping neighbours to dig out their sheep trapped in drifts; for nowhere do you find such genuine co-operation as between the best type of sheep farmers.

Many hundreds of sheep were lost, and many more sadly reduced in condition, which involves a whole cycle of troubles, weak lambs, no milk, and later in the season ravages from the green-bottle maggot-fly.

The shepherd made little progress towards recovery, so that at lambing time I found myself shepherd as well as foreman. I went each evening to his cottage for advice and then followed it as well as I was able; it proved an invaluable experience.

The shepherd was a man who will always remain in my memory: a simple Galloway shepherd, who had strayed a little from his native heath, who in the true sense of the word gave his life to the sheep. An ignorant fool to some, a morose old man to others, but finding that I never spared myself in looking after the flock, he opened his heart to me and taught much that has added pleasure, interest, and enlightenment to my life. From him I learned much of the age-old wisdom of the hills. Almost unlettered, yet with a mathematical mind worthy of a senior wrangler or famous astronomer, his calculations and observations of celestial phenomena, based on Celtic folklore (of which he had a great stock) held me spellbound. His simple laws, passed on *en bloc* twenty years later to an instructor in astro-navigation, may yet revolutionize the teaching of one of the most difficult branches of practical science.

Whole books have been written on standing stone circles by great and learned archaeologists without coming to any conclusions, yet this untutored peasant could demonstrate them for what they are, calendrical observatories. From his teaching I can calculate the time within a quarter of an hour at any season of the year by a glance at the stars, or the age of the moon in days for any date.

Marking the positions of Deneb, in the constellation of the Swan, at dawn on the day the rams were turned in with the flock, the shepherd expected the arrival of the first lambs when Vega occupied the same position at sunset. While other people, using gestation tables, roamed the hills at night looking for lambs, our man would remain tranquilly in his cottage, saving his energy until the stars showed that the lambs were due.

With the decline of physical strength his mental faculty became more acute. At no time did he confuse astronomy with astrology or other pseudonymous science, but his intuition was uncanny. One day the farmer's wife told me that as he could no longer look after himself he would have to be sent to an institution. Knowing it would kill him, I threatened to leave on the day he was sent, which compelled her to drop the project. The same evening, as I approached his cottage, I saw him standing on a little hillock before the door, leaning on his long crook and with a plaid over his shoulder. From here he could survey the valley and the heights beyond. As he turned to me, his long beard and his piercing eyes, burning bright with fever, gave him a druidical appearance. The setting sun shining through the clouds and throwing a halo round the great riven peak of Black Law behind him, completed the illusion. Before I could speak a word, he held up his hand.

'My son,' he said, 'for what you have done to-day you will receive your reward. All that on which you set your heart will be achieved. Those that work with you will prosper, and any who work against you will be cast down. You will spend the best years of your life in a fat and sinful land, yet whenever you set foot upon the hills my spirit will guide and comfort you.'

What could I say? Two years before I had been confirmed in a fashionable London church. A well-fed and comfortable-looking bishop, exuding wellbeing from every pore, had laid his hands upon my head and intoned his apostolic blessing; and in my heart I had felt what a farce it was. Yet here in the shadow of the great hills, from this unwashed, half-starved, but utterly sincere ascetic, who dwelling in the solitary places of the earth had found for himself many of the eternal truths, I seemed to receive something concrete; a real and driving force that would make his words come true.

Cheered and fortified by the old man's prophecy, I threw myself more wholeheartedly than ever into my work. The future might be foretold, but it could only be achieved by constant application and study, unremitting care and attention to detail.

We had a wonderful lambing season that year, the weather having relented after playing havoc with so many flocks in the district during the great blizzard. As the days lengthened I was working eighteen hours a day, directing the work on the enclosed land and shepherding the hill. I was completely happy and contented in my work, experiencing a sense of divine vocation which I have never lost in doing anything in connection with farming.

At last my employer made a partial recovery from his serious illness. We rode together round the farm and up on to the hill, now white with ewes and lambs in the spring sunlight. For once he found fault with nothing, being deeply moved to be out again and on his own mountainside, always known as 'the Hill'. I seized the opportunity of pressing the shepherd's case; and he quite agreed with me that the od man should be allowed to die where he had lived, if someone could be found to look after him. Contentedly we rode on. Suddenly, he said, 'George, how much money have you got?'

Wondering, I replied, 'About £60, sir'.

'Well,' said he quietly, 'I am prepared to take you into partnership without capital. I, like my shepherd, am finished on the hill, but your youth and strength, and the ability you have shown, convinces me that you can manage it for me.'

'You are very generous, sir, and I have done nothing to deserve it.'

'Maybe not,' came the reply, 'but you have been looking after this place as if it was your own, and I realize that you will not stop long in any paid position. For the little time I have left, money means very little to me, but the flock is my life's work. I can only give you half of all I've got, my wife must have the other when the time comes. But her share will be no good without someone who can show—has shown—the sheep come first, last, and all the time. Think it over and tell me on Sunday.'

A very tempting offer, this. No books were kept and no valuation made. Hill farmers are often so superstitious that they do not even count their sheep for fear there should be less next time. The bank book is the only record. If the balance is better than the previous year at the same date, all is well. If not, it cannot be remedied, except by work and prayer during the following year. A file of newspaper clippings giving the numbers and prices of ewes and lambs sold at the annual sales is the only indication of the value of the farm, the produce from the enclosed land being consumed on the spot, sufficient cattle and wool being sold to pay the rent and wages. If the farm changes hands the stock is valued by the neighbours, at considerably more than the market price, as an open hill farm is valueless without the stock which have become acclimatized over many generations to the particular mountainside.

As near as could be calculated the share offered would be worth about £2,100. £300 a week for doing what was only my duty by my employer and the flock during the seven weeks I had complete charge. On the other hand, so small is the gross return from this type of farming, my share would only bring in about £140 per annum. Also I could see no real future for improvement, the hill was steadily deteriorating for lack of cattle and wethers, which are no longer kept owing to the changing demand for young mutton and beef. To attempt re-seeding is to attract other flocks from the open sheep-walk and the poisoning of the fresh grazing, for a sheep's worst enemy is another sheep.

I could put up with the farmer, who since his illness had given up drinking, but I disliked his wife, and had in any case set my heart on going into partnership with my younger brother; and so in due course I turned the offer down.

Then for a few more months I carried on as usual. The great annual sheep sale came; it was my responsibility to get the flock safely to market. The master and shepherd would come later—it was their last sale. Our stock made a higher average and for a larger number than ever before. Congratulations poured in on every side, the other shepherds accepted me as one of themselves, and I felt it was one of the greatest days of my life.

On the strength of such a good sale, a neighbouring flockmaster made an offer to take over the farm to run with an adjoining property, and on pressure from his wife, and the assurance that the old shepherd could keep his cottage, my employer agreed to sell. So once more I was free, and with many expressions of goodwill, we parted.

After a short holiday, during which I endeavoured to catch up on my studies in the theory and science of farming, which I had neglected for want of even a spare half-hour at night, I obtained a situation in Oxfordshire. Here again I went for a month on trial. The job seemed too good to be true, as I was to be allowed to have a little land of my own on which I could keep some stock. Only the customary hours were worked, so that it would be possible to look after my animals early and late, and in the winter I would be permitted to fit it in with my ordinary work when the days were short. I felt I had learned how to *farm* in Derbyshire, with good stock and modern methods; how to do many odd jobs and make profits out-of sidelines in Essex; how to work and take responsibility amongst the rocks and heather of the north; but here I was taught how to *live*. Mine was a kindly and a considerate employer who ordered his life on Christian principles, a very great change from the hard-driving, hard-swearing, and hard-drinking farmers I had found elsewhere. Never seeking to take advantage of anyone, he was true and just in all his dealings; an excellent example to any young man who has to make his way in the world. His sound advice, and kind recommendations to merchants, auctioneers, and traders proved invaluable when we took Oathill Farm. From his wife too I received many kindnesses for which I wish to place on record my appreciation.

It was a typical Cotswold farm, mostly arable, and with a hurdle flock to maintain fertility. Milking cows had replaced fattening bullocks, but otherwise it was the old traditional system which had served English farming so well for many generations, but was breaking down under changing economic conditions. The old test that a farmer should get a 'rent' off his sheep, another off the corn, and a third off cattle, with which to pay one each to the landlord and labour, having the other for himself, no longer applied. Labour alone wanted two 'rents', and if the landlord had the other there was nothing left for the farmer. The only remedy would be to increase production with, say, pigs and poultry, as an extra section to balance increasing costs, and not to sacrifice a fine system which had proved its worth. The paltry expedients by which many farmers tried to meet the situation were exposed by an old labourer when he said to me, 'If it don't pay to do well, master, it can't pay to do bad'. I saw quite clearly the good qualities of the system, by which fertility is maintained, and the remedy for the faults, which I hoped to incorporate in our own methods when we were able to take a farm. Meanwhile, I spent a very happy and enjoyable year, learning to love the Cotswold country, which like its people was poor but honest and would respond to good treatment. The land was cheap to rent or buy compared with other districts, and so I asked a local auctioneer to find us something within our means, and with ample scope for youth and energy.

During the years I was learning farming my brother had not been idle. For two years after he left school he helped an elder brother, who was starting a new business in what was then a completely new industry with a great future. Building their own factory, designing new and wonderful machines, training labour, organizing an office system, and creating a demand, he had unique opportunities to develop his talents and earn money. Yet in spite of all the thrills of business building and a comfortable and assured future, the call of THE LAND was too strong, and so George and Frank Henderson, FARMERS! entered into partnership on the 4th day of March 1924.

# CHAPTER TWO

## THE PLAN

In the few weeks between signing the agreement and taking possession of the farm, our complete plan for farming the land and stocking it was carefully drawn up, the financial aspect also being carefully budgeted. It is interesting to look back now and see that through all the changing fortunes of agriculture it has never been necessary to change more than a few details of it. Working on the assumption that we would be able to live on the income from stock, using the return from corn sales for debt repayment and improvements, it was calculated that we could establish ourselves as tenant farmers in seven years and then buy the farm freehold in a similar period. The first step was achieved in five years and the second in another five, which indicates that our plan was not too ambitious.

We believed then, as we do now, that the greatest tragedy which had overtaken agriculture was not the repeal of the Corn Production Act, under which prices were to be maintained and linked to wages, but the Agricultural

Holdings Act under which tenant farmers were given freedom of cropping, or in other words freedom to rob their holdings, incidentally and inevitably robbing themselves in the long run, for the preservation of fertility is the first duty of all that live by the land. One hears a lot about the rules of good husbandry; there is only one—leave the land far better than you found it. In the soil lies all that remains of the work of countless generations of the dead. We hold this sacred trust, to maintain the fertility and pass it on unimpaired to the unborn generations to come. The farmer above all must have faith in the future, even the narrow demands of national extremity must not outweigh his judgement and justify the exhaustion of his farm, for a civilization lasts but a thousand years, while in his hands lies the destiny of all mankind.

In the most difficult years of depression a really good crop would clear expenses and cashed through stock would show a reasonable profit. Yet so many farmers tried to remedy the situation in which they found themselves by growing a bigger and bigger acreage of corn, with an inevitable smaller and smaller yield, to be sold at lower and lower prices as the years rolled by, until they were as insolvent as their land was impoverished; and in many cases farms became derelict. The solution, which appealed to some, of tumbling the land down to grass and farming with the proverbial 'sheep dog and a roll of netting', was also doomed to failure, because profit depends on intensity of production, and the stock they could carry was no more, and in many cases a great deal less, than could easily have been maintained on arable land. However, this system had the saving grace that it left Nature to preserve the inherent fertility of the soil, which the farmer showed himself incompetent to manage.

Even the National Farmers' Union at that time advocated a subsidy on arable land as a remedy on one hand, and advised their members to reduce wages on the other—which resulted in the Agricultural Wages Act—without any subsidy. In more recent years a subsidy has been advocated on wages—yet never once did they say, 'Make your farm so productive that you can afford to pay good wages, or teach your men to be so efficient that they can earn them'.

For us the wisdom of the ages was available. A hundred and fifty years of British farming had proved the value of rotational cropping, whereby fertility can be indefinitely maintained by the return in proper order of the manurial residues of the crops grown; and who were we to change it? A five-course rotation, fallow crop, corn, corn, ley, corn, had been common on the Cotswolds, and was well suited to our requirements and the layout of the fields.

In view of the dirty and neglected state of the arable a thorough cleaning of the fallow would have to be the basis of our farming. The destruction of couch grass, one of the most difficult and expensive weeds to eradicate, would be our chief problem. To follow the usual method of working out with many cultivations, collecting and burning, we believe to be the greatest mistake a farmer can make. This robs the land of fertility to a greater extent than the taking of a heavy crop of wheat, and leaves behind tiny pieces of root ready to grow again and befoul the field with this obnoxious weed. Ploughing in the weed every month from February to August will thoroughly exhaust and destroy it, and fertilize the land as to the equivalent of mustard grown for green manure. Our method, still believed by many farmers in the district to be wasteful and risky, was the basis of our rotation, yet on their farms we see the 'squitch' fires burning year after year, yet we have never had to destroy a handful by this slow and uncertain means. One-fifth of our land was to be cleaned each year by continuous ploughing. Coltsoot, another troublesome weed in our early years, was easily overcome by deep ploughing, or as deep as the nature of the soil would permit. Thistles, another bugbear of farming, were overcome by the same means and by planting winter corn. Docks are the only weed which continue to be a nuisance; a field will appear free from them for years, until a deeper ploughing germinates a full crop which can only be laboriously removed by pulling.

Once cleaned a field would grow two good crops of corn if generously helped with artificial manure. The use of artificials is only justified by the intention of making a bigger and bigger manure heap. To sell off corn and straw grown by chemical manures should be made an indictable offence.

'Seeds', the common term by which farmers describe the mixture of clover and temporary grasses, and the basis of fertility in rotational farming, was very difficult to establish in our early years, but havy manuring with farmyard dung has brought about such an improvement that a good crop of self-sown clover can be depended upon to appear on any stubble in a wet autumn.

After clover, penned off or dunged, a fair crop could be grown, but here again artificial manure justified its use to provide more corn and straw to be consumed on the holding.

Yields have been carefully checked and tabulated over the years, and have justified our belief in the proper use of every source of available manure. Superphosphate and sulphate of ammonia have given satisfactory results every year except one. Potash has not justified its use, from the point of view of increased yield (though may be useful in maintaining a balance), once a field has been heavily dunged in the rotation.

Wise and thoughtful people in recent years have drawn attention to the misuse of artificial manures. In our experience they do no permanent harm if used in conjunction with farmyard manure and ploughed-in green crops. The health and stamina of stock and crops is not affected as long as the balance is maintained and the cycle continued of—better crops—more stock—more manure—more humus—and more corn to be consumed on the holding.

Once a field had been cleaned of couch by thorough bare fallowing then roots could take their place in rotation and add to the stock-carrying capacity of the farm. It is hopeless to plant even potatoes, which are considered a cleaning crop, where couch abounds, for its spreading roots will penetrate right through the tubers themselves (and also, as an old gardener told me when I was very young, right through a man's heart, but perhaps he was speaking metaphorically).

Our livestock scheme was more ambitious, if on a modest scale. After paying the small ingoing of tenant right valuation, and taking over the dead stock and machinery which was sufficient to work the farm, only £200 was available, and this at a high rate of interest. Yet nevertheless we determined to have nothing but pedigree stock. For others, the risks, dangers, and disappointments of the open market; in which one as often buys disease, vice, and trouble as healthy and profitable stock. For us, only the best would be good enough. We could not buy it, so we would have to breed it. We therefore allotted £50 for each section, poultry, cattle, sheep, and pigs. Each branch would have to develop out of profits. In this way our experience would always be equal to the stock we had to manage, without the temptation to invest capital too heavily on a boom in one section, and lose it in the next slump.

Hedge-laying and tree felling were also carefully considered. There were a hundred chains on the farm, and all except seven were badly neglected. We sold £60 worth of firewood out of them in the first year, which indicates the state in which we took them over. Surface roots spread a chain into the fields, robbing the crops and breaking the implements.

The buildings would be reconstructed to accommodate the stock, and a road built, when time and money were available.

Labour presented little difficulty. We decided to employ one man for the first year, to help with the hedging and ditching, until the younger partner had gained practical experience. The man's wages could be more than half met by the sale of firewood.

For ourselves, we adopted the principle 'the world forgetting, by the world forgot'. We intended to work about eighty hours a week, or twice the output of the ordinary labourer, allowing for the fact that twenty per cent of their time is wasted for want of careful planning and real interest in the job. If it were not, they would not be labourers very long. On the other hand we proposed to live on half a labourer's wage, which is quite easy if a good part of one's food is produced on the farm. This method can be recommended to anyone who cares to try it. One acquires the serenity of outlook only usually found in a monastery. One is not troubled with second-hand opinions absorbed from the daily paper. In fact for the first five years we did not buy one.

It may be that man is intended for higher things than looking after lower animals, but we were happy and contented, with simple living, the health and vitality of youth, our plans and dreams, being our own masters, and serving the LAND. In the hours when other young men of our class were shooting, playing cards and tennis, or taking a girl to the pictures or on the river, we were working. For everything one has in this world a price has to be paid. The hours we put in then have paid substantial dividends ever since, even if we missed some of the love, light, and laughter which is the prerogative of youth.

To some this book will appear egoistical and boastful, and the writer regrets he cannot strike the modest note that characterizes the work of the best farmer-authors. But it should be remembered that we have always lacked the courage to launch out in anything that was not backed by sound principles and of which we had a thorough grasp. It is said that even a fool can learn by experience; we have always preferred to learn from others. Once when I was a pupil and helping to cut in half a very hot hay-rick, which had heated almost to the point of spontaneous combustion, and sweat was blinding me and soaking my clothes, a farmer came along. After watching for a few minutes, he said, 'You are a very lucky young man'. 'Why, sir?' I inquired. 'Because you are getting this experience for nothing; someone else is paying for it.' How true. Needless to say we have never had a hot rick on our farm. It is the same with almost everything; we studied, compared, and observed before attempting it. Somewhere there is always someone who is doing a job a little better and there are many who are doing it a great deal worse; from either a lot can be learned. Now if we were to set up in business as art critics or designers of ladies' underwear, only then could a book be written on our ludicrous adventures in a business of which we know nothing at all. To us farming is a very serious occupation, and successful farmers, like millionaires, seldom smile. The lighthearted manner in which authors describe being thrown at every jump, or funking it altogether, speaks well for their courage. I was taught by an expert how to present a sound, trained, balanced, and collected horse knowing exactly what it could do, otherwise I would never have had the intestinal fortitude, spelled with a capital G, to attempt it. It is very much the same with our farming.

# CHAPTER THREE

## THE POULTRY

The poultry section, the most important branch of our farming both from the point of view of profit and of the maintenance of fertility, has been built up from the very smallest beginnings.

In the autumn of 1914 we were asked to look after a small pen of Light Sussex for a gentleman who wished to go and help clear up an odd spot of trouble on the Continent. He expected, in common with many thousands of his fellow countrymen at that time, that it would all be over in a few weeks or months.

This we were quite happy to do, without also realizing that we would continue day by day for nearly five years before our trust would be completed. Every day we fed and watered the birds, collected the eggs, entering the number in a book, selling them each week. The cash account showed the sales of eggs, and in the opposite column the expenditure on feeding-stuffs. Thus at the early age of eight and ten respectively we started a system of book-keeping which in the course of time would grow into the carefully analysed accounts which to-day enable us to show the cost of producing anything on our farm from a day-old chicken to a fully trained farmer over a period of twenty years.

By the spring of 1915 it was disclosed that Lord Kitchener had prepared for three years' war, so to maintain our trust we set two broody hens, reared the chickens, sold the surplus birds, and banked the profits.

It is interesting now to look at the old record, in which childish handwriting made such entries as '19 May 1918, Poultry not cleaned out till afternoon, daylight raid preventing in morning.'

Apparently we took our work very seriously, regretting that the birds were not cleaned out on Saturday morning. Another entry: 'Eggs 7d. each. Retailed by shop at 9d. same evening. Profiteers! Retail our own in future.'

Cockerels made twenty-five shillings each at the top of the market. A striking contrast from the controlled prices in the Greater War of 1939.

And so month by month and year by year we carried out our trust until 1919, when a somewhat battle-scarred warrior returned to civil life, and we expected to return the stock and hand over the profits. The accounts were duly presented, and carefully scrutinized by the owner.

'You haven't charged anything for labour,' he said.

'No,' said the junior partner, 'it was a pleasure to look after them for you.'

Did the senior partner detect a tear in the eye of the man who had faced death, disease, and disaster, in carrying out his duty for over four years? Strange, but true. Somewhat huskily, he made over the birds and profits for our trouble.

Never was youth so richly endowed! The stock had grown from six birds to twenty-seven, a good strain of real old-fashioned Light Sussex, originally obtained from one of the founders of the Light Sussex Club in 1904. They had been very closely interbred, but any lethal mutants must have been bred out, as no serious fault ever appeared. Big, fine birds, not very good layers as judged by modern standards, but with health, vigour, and longevity, for which in recent years many poultry breeders have searched in vain, after breeding for egg production and laying-test winners. It is our carefully considered opinion that more harm has been done to the poultry industry by the trap-nest than by any other modern innovation.

The junior partner looked after the flock, while the elder was away learning farming. And then, when we took the farm, the birds were brought down with the furniture, to lay the foundation of the stock, which in later years has been sold by the thousand, some birds going as far north as Leningrad, and others as far south as Ceylon.

In 1925 two sittings of eggs were brought from the strains of two of the most successful breeders of their day, Mr. Marcus Slade, and Mr. Rossal-Sandford. This blood incorporated with our own stock increased egg production, without loss of size or stamina, by nearly fifty per cent. Only once since have we introduced new blood, and that from the original Rothschild strain, kept pure by that very capable and successful breeder over a great many years, Captain Coates, of Kilworth, Leicestershire, grandson of the founder of the first Herd Book; and who could know more of the value of pedigree breeding? His birds matched ours in every point we value. Over the years, of course, we have tried stock from other breeders, but in no case have we found fresh blood worthy of incorporation in our strain. It may be interesting to note that after thirty years of line breeding, the chickens rear and thrive as well from the pure pens as the sex-linked cross with Rhode Island Reds from which one would expect hybrid vigour.

As from 1926 we have had as many as ten separate pens, the cockerels used never being closely related to the hens, although of the same strain. How often have we noticed the poultry farmer buying trouble who thinks he must have fresh blood. Breed close and cull hard is a sound principle in practice, and is supported now by the findings of the leading scientists in the field of

Mendelism and genetics. In our experience you can breed anything you like in five generations by careful selection, and ruin it in one by haphazard mating, or careless management.

For the first year at Oathill we reared as many chickens in the spring as our limited resources would permit, and the number of broody hens available. One hundred and ninety-nine out of two hundred and one hatched. How they revelled in the sunshine and enjoyed the fresh grass and insect life. What a change from how we had reared their mothers and grandmothers, cooped up on a tiny space polluted by the smoke and grime of London. In that year too we were free from the depredations of ground and winged vermin, which are the most annoying of all losses because it is the finest and best birds which are always taken.

Of the two ground vermin are the worst, because they kill simply for the sake of killing, whereas birds of prey only kill what they can eat or carry away. Poultry are never safe, even at night, from foxes, rats, stoats, and weasels, unless suitable precautions are taken to protect them from these pests.

Good houses, which are closed regularly as soon as the birds have gone to roost in the evening, are essential. Before all our large poultry houses we have an ordinary steel door-scraper mat, made from flat bars on edge, up which no fox has ever ventured; it even puzzles the farm cats or dog to negotiate them until they learn to jump right over. It also keeps out lambs or pigs, who are tempted to enter for shelter or in search of food.

Small-mesh wire netting or shutters are fitted to all open-fronted houses, because rats, stoats, and weasels will sometimes manage to squeeze through ordinary two-inch mesh netting which is often fitted to poultry houses. Although they seldom attack adult birds, they will kill scores of young chickens, which may have just been moved from the brooder houses, if they are able to gain entry through the smallest hole. Occasionally we get a rat which will kill hens at night; it is one which has invariably taken up its quarters somewhere inside the house, and it has to be found and destroyed at all costs. If a rat only tears a hen badly, the other birds will resort to cannibalism and finish it off.

In guarding against stoats and weasels it must always be remembered that it is possible to pull a dead stoat through your thumb and finger when the tips are pressed together, and it will be realized through what an incredibly small hole these destructive animals can squeeze.

There are many well-known methods of killing rats, all of which should be employed where possible; well-trained cats usually prove to be the most effective and persistent means of destruction. The stock at Oathill usually consists of about a dozen, all descendants of a cat we brought with us twenty years ago. Poisoning, in places where the cats cannot reach, together with trapping and snaring, has kept us reasonably free, although we occasionally get a plague of them, if the larger arable farmers in the district leave their corn ricks too long and breed a few hundred to be let loose when the corn is threshed. We have seen them running across the fields in droves.

With stoats and weasels cats are unsatisfactory, as they object to the offensive odour, and seldom attack. If chickens are killed in the open by these animals a little patience will usually be rewarded if one waits with a gun, as they soon return to the scene of their kill, and appear to have little fear of human beings, apart from actual movement. If there are several stoats and weasels about (and the appearance of one of these animals is usually a sign that there are others in the vicinity), a dead chicken, suspended over a carefully set trap, will usually prove an effective bait. Unlike rats, stoats and weasels are nearly always on the move, so that if no losses are experienced for three days, it is fairly certain that the marauder has passed on.

Foxes are the most serious menace with adult birds, and what is most annoying to the poultry-keeper is the fact that if there were no hunts there would be no foxes. In countries like Holland, where the farmers will not tolerate them, the appearance of a single specimen is reported in the newspapers. But I do think that the majority of losses are due to the neglect and carelessness of a few farmers who teach foxes to take poultry by failing to shut their birds up directly they have gone to roost. The fox renders some service to agriculture by destroying rabbits, and we always regret having to kill one out of the way. We have lost about forty birds in twenty years, but to maintain such an average means that every fox that takes a fowl signs his death warrant. If he only runs through the poultry he is left alone. They are not nearly so difficult to destroy as is generally supposed; their proverbial cunning is ill matched against the patience and knowledge of man. But with these supersensitive animals it is a sound rule never to attempt to shoot, trap, poison, or snare as a single remedy, but to employ every possible method from the first and at the same time.

It is a remarkable fact that a fox, especially a vixen hunting for her cubs, will inspect a dead fowl suspended over a trap set in a deep ditch and pass on to a poisoned bird, examine it carefully from every direction, and finding no trap take the bait without suspicion. Or going the other way and suspecting poison in the first will go on and blunder into a trap in an attempt to pull down the bird in which no poison can be detected. On the other hand, if both trap and bait are avoided, a wire loop set six inches above the ground in a suitable hole in the hedge may prove that many methods justify the trouble taken.

That we have been able to observe this many times is due to our waiting in a suitable position with a gun, and the fox having been detected while still out of range, has had the incentive to take a hurried meal which has been so thoughtfully provided elsewhere.

At the same time, while employing methods which can be devised, I have shot foxes at less than forty yards' range, with no more cover than lying down in a patch of rough grass in the middle of a large field.

In many cases a rabbit is a more suitable bait than a dead fowl for trapping. For poisoning a rook can be recommended, as the strong smell tends to disguise the preparation used. A few grains of cyanide of potassium is all that is necessary, and domestic animals do not tend to eat these birds, even if given the opportunity, while foxes frequently do under natural conditions.

Farmers who are reluctant to use drastic methods, or who find that poison is not safe, would find a two- or three-section poultry coop or fattening crate a satisfactory trap. Place a suitable bait in the end section and close the slide. The middle sliding bar should be made to run very freely, weighted, and supported by a light twig which will be brushed out as the fox enters, dropping the sliding bar on the trap-nest principle.

A log of wood supported on a Y-shaped stick has been used successfully, where entry has been made under wire-netting, but it is rather on the hit-and-miss principle. A better means, but requiring a little more ingenuity, is a bent-down sapling with a wire loop, the springing of which pulls the fox right off the ground when the trap is sprung. Sufficient to lift eighteen pounds is necessary. If it is only desired to scare foxes away from poultry pens, tin cans fixed to the wire-netting in such a way that they rattle in the wind, or at the slightest touch, will often prove effective. Probably the finest deterrent of all is an electric fence, fixed six inches off the ground in a narrow track cut through the grass. These highly strung, sensitive animals do not come back for a second shock. I have been told they jump six feet off the ground when they touch the wire, and it is certain that a fox carrying away a hen cannot negotiate it without getting direct contact if this method is used, although it is not applicable to all circumstances.

With hawks and crows, shooting is the best method. Rooks are sometimes blamed, but in our experience they do not kill chickens although they will clear up a carcase left by a carrion crow. They are easily distinguished; a rook has a bare patch on each side of its face, while a crow is feathered to the beak. It is not everyone who can spare time in a busy season for bird scaring so other methods have to be adopted. One or two broody hens running loose will drive off hawks. Also it should be remembered that carrions only take chickens for their young, so that if the day on which the first chicken is taken is noted, and the location of the crows' nest determined, one has only to go there twenty-one days later, which is the day when the young climb out of the nest, and they can be shot without difficulty. The nest itself is shot-proof, being lined with a foot of sheeps' wool, so that it is no use shooting at it. The belief that carrions only build in unclimbable trees is erroneous; my brother has destroyed a number by climbing, although he has always maintained that no tree is unclimbable that will bear his weight, thereby justifying the Darwinian theory from his earliest youth. Those who favour the Lamarchian or the orthogenesis theory of evolution will contend that it is better to wait three weeks and shoot the young crows! In passing, it might be mentioned that we regard rooks as entirely beneficial birds; like poultry they have to be kept off freshly sown corn and the matured crop at harvest; for the rest of the year they render a valuable service in maintaining the balance of nature.

In our first year also we were able to buy Indian Runner duck-eggs very cheaply and reared a nice stock, with our ever-running stream, and two or three acres of swampy ground proved ideal for them. When they came into lay we regularly got forty-nine eggs from fifty birds, and the worse the weather the better they laid. Mated with drakes obtained from Mr. Reginald Appleyard of Ixworth, Suffolk, one of the most successful breeders in the country, we did very well with them indeed, and only gave them up when an aerodrome was re-established a few miles away. Low-flying planes terrified them and, causing them to moult out of season, stopped them laying. Unlike hens, which quickly adapt themselves, even the second or third generation of ducks would be put off laying by a single plane hedge-hopping, which was a favourite occupation of the R.A.F. in those days.

With geese we were even more successful. With a lot of rough grass and weeds for them to graze, the small quantities of concentrated feeding-stuffs required, and the low initial cost of the breeding stock, they were ideal for a semi-derelict farm and our limited financial resources. A goose egg for sixpence in April, and a fattened goose at fifteen shillings to a pound at Christmas, gives the quick turnover of capital which is essential when one has to farm, live, and earn capital at the same time. Yet few farmers seemed to recognize their possibilities. Three geese and one gander will, with good luck, produce one hundred goslings. The initial cost of the stock birds will be less than that of a yearling beast; but the hundred goslings would sell for £50 at the end of the season over and above the cost of the meal and grain they require. But in the case of geese it must not be forgotten that the stock birds remain at the end of the year ready for another season. With good management they continue to lay fertile eggs for many years; some we sold in 1926 were still giving satisfactory results fifteen years later.

The most common breeds kept in this country are the Emden and the Toulouse, the latter being the most suitable for our purpose in producing large numbers to consume surplus grass, and as we were prepared to hatch in incubators and rear with hens. The Emdens do not lay so many eggs, although they are better sitters. Both breeds, however, are equal for table purposes.

Any airy shed is suitable for housing the birds, providing it is fox-and weather-proof, and well littered down with dry straw. Dummy eggs should be placed in the shed, or the geese will lay their eggs about the farm. Swimming water is necessary or the eggs will be infertile. As they are immune to all, the common poultry diseases, it is extremely unusual to lose any geese, other than by accident, provided the birds have sufficient exercise, and this they will always find on free range.

We hatched them in ordinary hot-air incubators, run at two degrees lower than for hens' eggs, spraying the eggs daily with warm water. Our large Light Sussex broody hens could rear six goslings apiece, because they require very little heat after the first ten days. The period over which they require brooding at all varies from a fortnight to three weeks according to the season of the year.

Soft food is essential for the first month, after which they can find their own living, providing there is plenty of fresh young grass. Some shelter in very hot weather may be necessary. From the beginning of November we fattened them generously on barley meal and skimmed milk, and they never failed to leave a useful profit when sold at the Christmas markets, realizing far more per pound than the finest fat cattle at that time.

What a lovely sight it is to see a flock of say a hundred geese come flying in from the stubbles when called for their evening meal. They used to turn into the wind, or downhill, run a few paces, take off, and sweep into the farm, on more than one occasion taking down the wireless aerial as they came over the house. When I went to plough they would trail out in a long line behind me; as they could not walk far or on the turned furrows, they would sit along the unploughed land, draw back as the horses passed and then drag out the long couch grass roots and worms, each goose on its own little stretch of perhaps a couple of yards.

In those early days, too, we planted mustard for green manuring, and drilled a couple of pounds of trefoil clover with each bushel of oats in the spring, so that after harvest the geese had masses of green food, and any surplus could be ploughed in.

After a few years geese were replaced by sheep and then by cattle, which leave a better margin of profit per acre once the capital is available; but for cheap stocking and land reclamation geese have much to recommend them, in fact we would prefer them to ordinary commercial cattle rearing. But even four geese to the acre, leaving a net profit of £3, cannot compare with well-managed sheep or pedigree cattle. While geese could clear up clover, mustard, or stubble, they cannot consume roots or tread straw into manure, which is so desirable if the land is to be brought into good heart; and for that reason we regarded them merely as an expedient for the times and not as an integral part of our farming system, useful and attractive as they undoubtedly were as a sideline until we had drained the swamps and cleaned the land.

No other branch of our farming compared with the pure Light Sussex poultry. From our 199 chickens we reared 100 good pullets, which together with our adult birds numbered 123. By November they were in full lay, and with eggs generally scarce and dear, were clearing over £4 a week after the food had been paid for. This seemed too good to last, when one thought of the meagre returns from ordinary farming at that time, or compared with the ten shillings a week over and above the cost of board and lodging which I had been earning the previous year, working on a farm. But my brother was not impressed. He pointed out that this was only equal to a shilling per acre per week over the farm, and we would need a thousand good birds before the farm was a real business proposition. With his training in London in modern methods of business, he could not agree that a farmer must be prepared to accept small returns and a low standard of living for the privilege of working on the land. He regarded us as potential £1,000-a-year men, having studied those who earn it in the cities, and if my knowledge and technical efficiency was such that we could produce eggs, when others obviously could not or eggs would not be so dear, then all would be well.

Unfortunately it requires £1 a head to capitalize a thousand head poultry farm, while we were building from £50, and some of our profits would have to help keep the farm and us, besides building up the stock. However, time and youth was on our side, and all comes to him who waits.

In the second year we again reared as many chickens as possible, the numbers being dependent on our limited resources and the number of broody hens available. Later in the spring we reared more ducklings and geese, so by autumn we had a nice stock of birds, and made a contract with a London shop for selling eggs, as at this stage we were not in a position to claim to be poultry breeders, and to do better by selling hatching eggs and stock birds.

Late in the year we bought a very old second-hand Ford lorry for a ten-pound note, and this rattled the seventy-odd miles to London every week, and continued to do so for nearly two years without ever letting us down. By starting before 5 a.m. my brother was back by midday, and we thought the half-day was well justified in maintaining the personal touch, which is so important in business, besides saving the heavy railway carriage, and enabling us to sell surplus cockerels, ducks, and geese to the best advantage. At that time the difference in price between the local markets and in the metropolis for eggs was 4d. to 6d. a dozen, and cockerels 8d. to 10d. a pound. The lorry also enabled us to sell produce for our friends and neighbours at good prices while also earning carriage. Occasionally small return loads were found in London for Oxford and district, for which we could quote cheap rates, when every shilling was welcome.

Then the shopkeeper with whom we had been doing business every week to our mutual satisfaction sold out to a large combine, the contract being taken over. What a difference we found! Every week there were deductions for cracks, blood-spots, and grading. Careful checking and candling at this end brought about no improvement. So waiting our opportunity when eggs were scarce and dear in the autumn, we held up delivery for two days and then took them and said we wanted to see the eggs tested and graded. No facilities were available on the premises, but the eggs were put straight into bags, first and second grade the same, and sent straight out to the waiting delivery vans. Our cheque, from which the usual deductions were made, about ten per cent, came along in a few days. We thereupon refused to continue delivery and broke the contract. The firm were very indignant and even threatened to sue us for breach of contract and libel, but nothing came of it. So much for Big Business—give us the small, honest trader every time. At that time of the year we could sell all our produce without difficulty and we had other plans for the spring.

*EGGS, FLESH, AND BEAUTY COMBINED*

*JERSEYS WINTERING OUT, 1940*

During the great General Strike in May 1926 we had a good stroke of luck. We went to a large poultry farm sale, and as the printing trade was on strike no catalogues were available, and the auctioneer was compelled to describe each lot as he came to it and sell it accordingly. We bought several small items, and then they came to the poultry. Sex-linking, the method of mating a red cockerel on a silver pullet or hen giving in the progeny white cockerels and red pullets when hatched and providing a simple means of detecting the sexes, was not then well understood. We had read Professor Punnet's book on the subject and were familiar with the process; but the auctioneer proceeded to give a lecture on sex-linkage, Mendelism, and genetics, which was largely wasted on the assembled farmers, and then he proceeded to offer the cockerels as pullets and the pullets as cockerels. This was our opportunity. We bought three hundred so-called cockerels, actually pullets, for an average of Is. 8d., each leaving alone the so-called 'pullets', which realized six or seven shillings each. What the poor buyers must have said when they all grew into cockerels can only be left to the imagination. The auctioneer's conditions of sale covered 'all errors and descriptions' so there would be no comeback. We had a customer in view for our three hundred pullets at home, and they were duly delivered and we netted a modest £50 for our day's work. Take a hint, gentle reader, always be well up with the latest scientific developments in farming! It was perhaps a little mean on our part to take advantage in this way, but earlier in the sale I had respectfully drawn the auctioneer's attention to some other point in both the buyer's and seller's interest, and was sharply told that if I had come to crab the sale, I had better go away. So it is unlikely that I would have been believed. On the other hand, if it is thought we took too big a profit from our customer, it may be said he sold them again at a substantial increase immediately.

Our profit we invested in three large second-hand incubators and six hovers for rearing chickens, as we had decided to give up selling eating eggs and to concentrate on hatching and rearing for stock and table. As Light Sussex was the premier breed for fattening, and in demand all the year round, it was a practical proposition.

At the same time it was a very decisive step in view of our very limited financial resources; and we would sadly miss our small but regular income from the eating eggs. If we were unable to sell the day-old chickens when hatched, we would be unable to afford the food they required and thus make a double loss. The only method open to us was skilfully worked credit, because our bank manager had very definitely declined to assist us in any way.

This little problem I could safely leave to our tame economist, while I devoted my attention to plough, for it must be remembered that at all times we also had our other work to attend to—the land, with its weeds and chain-wide hedges, our few sheep, cattle, pigs, and horses. But by this time my brother had become quite a skilful poultryman, besides compiling some very valuable data showing percentages of eggs we could expect to hatch, quantity of food required up to any given age or weight. But more important of all, he had made very valuable contacts with two or three corn merchants, who not only liked, but trusted him. Looking back over the years, I am inclined to think that although my four years' training enabled me to bring technical knowledge and skill to our partnership, it was his business acumen and organization which enabled us to start on a sound foundation. I think he would give me full marks for the general outline of our plans and principles on which the farm was to be stocked and worked, which I had so laboriously acquired from the wisdom of old and experienced farmers and the study of agricultural science, and which he accepted without question, but he so adapted, adopted, and improved them, it would be like comparing an uncut diamond with the polished stone.

This was his plan. First we advertised hatching eggs, charging double the price of eating eggs, and accepting orders up to half our output. So that for all intents and purposes we had the eggs we retained free, apart from packing and carriage. Then we advertised day-old chickens at £5 a hundred, and booked orders up to half the number we anticipated hatching, so that for every chicken sold one could be retained for rearing. The price realized for the chickens sold, enabled us to pay for the food to rear the chickens we kept until they were eight weeks old: After that age comes the expensive time for running on and fattening. At this stage we were careful to order the feeding-stuffs so that they would be delivered on the first of the month; thus with a monthly account with the corn merchant the statement would not be received until early in the next month, and so by selling the birds in the third week of that month, when they were fifteen weeks old, we had the money to settle the account and take the discount by the last day. In this way the merchant financed the business to the extent of two shillings per bird—or about £12 per week on 120 chickens, which represented twice the capital we ourselves had invested in this section. Had we ordered the food haphazard or late in the month, with the result that the account would be received within a few days, we would lose half the credit and also the discount. The whole system was of course dependent on hatching on the right date and knowing the exact quantity of food which would be consumed over each monthly period, different-aged chickens requiring varying amounts as they grew.

From our experience we would say, few farmers work credit to the best advantage. It has been described as 'the life-blood of industry', if it is, then regular payments are the heart-beats. Merchants do not mind a slow payer, providing he is a regular one. It is a great mistake to pay a little on account when pressed, it is far better to clear an account month by month, even several months in arrears if necessary. At the same time it is very bad business to lose a discount. Five per cent lost monthly on £100 worth of business represents £60 per annum. Though we lived on the edge of a financial precipice for the first five years we were farming, we always had a balance at the bank, if only a few shillings; at the same time we never lost a discount.

In theory it is not the merchant's business to finance farming, but the bank's. Unfortunately in the days of depression they were very reluctant to do so, or so they told 'young and inexperienced boys', while losing vast sums on the old and experienced farmers, who were waiting for the Government to do something and thinking in terms of 1913 fifteen years later.

However, I digress. Profits in poultry fattening were small, about one shilling per bird nett, but the numbers handled and the quick turnover made it well worth while. Soon our old Ford was trundling five hundredweight of fattened poultry to London every week. As we had skimmed milk available from the cows we could finish the birds to the best advantage and made a valuable contract with a famous restaurant in Bond Street. While other poultry farmers were dumping poultry in glutting markets, or getting lower and lower returns from the commission salesmen in Smithfield, we were sure of our price even before the chickens were hatched, providing we could rear them.

With a good contract running we were able to sell day-old chickens which were surplus to our own brooder capacity and buy them back from our customers at a fair price to fatten, which was to our mutual advantage. The customer had a fair profit on the rearing, we had two, one on the day-old, and another on the fattened bird.

The low prices in the local markets tempted us to buy these second-class birds, for nothing could turn badly bred and half-starved, stubble-running cockerels into the best quality, and these we sold to ordinary retail shopkeepers on the route into London, whose customers could not afford prime chickens. There was, of course, considerable risk of introducing disease from stock which was offered by auction, but this we felt could be minimized by keeping special crates for them, and arranging our journeys so that they were not unpacked or fed on the farm, but went straight into town. In practice this system proved sound and we did not experience any trouble.

But first we had to teach ourselves to judge weights accurately. As we packed our own birds from the fattening crates, one would estimate the weight of each bird handled and the other would check it on the scale. In this way we soon learned to value any bird within a penny, which was approximately the price we could realize per ounce.

Then we went to market. There the trade was largely in the hands of a ring of buyers from the East End, who robbed the farmers of their just return by a pernicious system of one man buying the lot, loading them up and going to a public-house yard and then holding a knock-out auction, in which the difference between cost and price paid by members of the ring was shared among them. Any strange buyer in the open market would be run up to such a price that he could not come again. We, of course, were quite familiar with this, and laid our plans accordingly.

We arrived early, and went carefully round the pens and valued each lot. When the sale was due to start we returned.

The real value of perhaps the first lot would be four shillings. The auctioneer, from bitter experience, would ask say, half a crown, and be offered one-and-six by the leader of the ring, all the others crowding round so that no-one else could see what was being sold. The auctioneer would then ask one-and-sevenpence, the owner of the birds would perhaps bid in the hopes of helping up the price: The ring would all jeer in unison, but the auctioneer would take it, and then my brother or I would start and run up the price, stopping at our limit. If the East End traders went any further it was left to them at a price they could not afford. If they stopped in time, we had it at a price at which we could make a profit. This shook them badly. If they crowded me away from the pens, the birds would be knocked down to my brother or vice versa. It did not matter whether we saw any particular lot or not, providing the auctioneer announced the number, as we had valued them before. Sometimes we confused the issue by bidding loudly against each other, but always stopping at our limit, so that the auctioneer said 'Henderson', from whichever side of the crowd the bid came. This was too much for the ring, and we helped ourselves to quite a lot of pens at far below our price.

By the next week they had thought out a good plan to get rid of 'Henderson'. They located my brother by the name on the lorry when he arrived at the market, and offered him ten shillings to go down to the station and meet the one o'clock train to see if there were any crates for them, which would mean he would be away when the auction was in progress. He wanted a pound cash down, and this they paid. Meanwhile I had arranged with the auctioneer that I was bidding whenever I stood with my hands on the lapels of my coat. By these means I bought pen after pen without removing my hands. The ring ran up the price a little way, and cheered each time the auctioneer said 'Sold', thinking of course the birds were being bought in by the owners at a reserved price, as no name was announced. Never was business so brisk, and by the time my brother returned to announce that there were no crates on the train, the auction was finished, and we proceeded to pack the birds under the eyes of the astonished ring.

'Oi, Oi, Oi,' said the leader, 'vot's this?' 'Just a little carting job for this chap', cheerfully replied my brother, indicating me.

We amused ourselves thinking out new tactics each week, but soon came to the conclusion that we were wasting our time, for the little we could earn. The quality of the birds deteriorated week by week. General farmers never feed them well enough, thinking they can find all they need on the stubbles, and some weeks there would be so few on offer that the journey was not justified. We grudged also the time spent, although it made a pleasant break for an hour or two once a week, especially if it was too wet to be working on the land. It was nice too when the first pen was knocked down to us, to hear someone in the crowd of farmers say, 'Thank goodness! Henderson is buying', which ensured that something approaching a fair price would be obtained. The ring system was thoroughly bad, and was rapidly dying out in poultry auctions by the time Lord Darling's act was passed making it illegal. Every little buyer was forced into it, and would perhaps travel all the way from London, riding on the step of a lorry in bitter weather, to draw perhaps ten shillings in the knock-out auction. We felt sorry for many of them individually, and may say in passing that we have found the Jews as straight and honest to do business with as any other race or creed with whom we have come in contact.

For about eight months of the year pullets were fattened the same as cockerels. They do not make the same weight for age, but the food consumed per live weight gained is identical. Pullets hatched in the spring could be sold for laying, and we steadily built up a good connection among general farmers, who had not the time or convenience for rearing their own, or realized that it is far cheaper in the long run to buy well-reared stock at an age when they required no special attention.

Each year we kept back sufficient good pullets to enable us to increase our breeding stock by a hundred birds. At no time were we tempted to rush into poultry farming on a large scale, but intended to build it into our system of general farming, and avoid the pitfalls of the specialized farm, with its unbalanced labour, unnatural conditions, and the bogy of disease ever present through overstocking of the land. Never more than one hundred birds to the acre has been our rule for breeding stock, and for part of the year these have a run on stubble. With the pullets being reared they move steadily over fresh grass land or leys in fold units day by day, which confers the greatest possible benefit to the land from the manure they leave behind, while ensuring the health of the birds. This system has enabled us to increase the grazing capacity of our grass by three times, as judged by the sheep and cattle which can now be carried.

It is perhaps interesting that with all the new innovations in poultry-keeping we have never found it necessary or desirable to change a detail in our methods of management, housing, or feeding. True, slight variations have been made in the composition of poultry mashes from time to time, but this has only been for economy when certain feeding-stuffs were cheap, or too dear in comparison with others of equal feeding value. Our original mixtures give as good results to-day as they did twenty years ago.

Someone might say, what about cod liver oil, and intensive rearing which has only come about this last ten or twelve years? The answer is that we have always used intensive rearing for the first few weeks, even before it was recognized as a system, and we used cod liver oil meal, prepared as such under a trade name, empirically it is true, without knowing the scientific reasons for its value in combating the disorders arising from lack of sunlight and vitamin D.

It is only the scale of our operations which has changed. We have always reared in small units of fifty to seventy chickens under Hannaford 'Pioneer' hovers, under which the chickens do not come in contact with the lamp, its fumes, or its brightness. Quite a lot of care is necessary during the first few days, teaching the chickens that it is warm under the hover and cold outside; once they have learned they are no more trouble, and rear far better than by any other system we have seen, with their attendant dangers of fumes and fire, to say nothing of freezing or baking the chickens, or undermining their health in the even temperature of the battery brooder. In our early days these hovers were used in small, well-built 9 feet by 5 feet houses, but once the value of the system was proved we built a large insulated brooder house, to hold two thousand chickens, but still using the same hovers in similar-sized pens within the house. With vita-glass front, which can open right up, and roof lighting, it was so planned that every pen can receive direct sunlight every day and at all seasons of the year, when the sun is shining. With a high roof, double insulated floors and walls, air-extracting ventilators, and other devices, it is possible to rear good chickens with thirty degrees of frost, or in a heat wave. In this house the hover temperatures are kept at 85° and the floor temperature at 50°.

The chickens in this house are fed four times a day, and have the bucket type water fountains which are filled every morning. This chicken rearing has become the most factory-like of all our farming processes. With everything standardized in each pen, one has only to glance down the house to see if anything is amiss. The birds are regularly culled and graded as they grow, and are moved out in even numbers to the rearing ground at six to eight weeks old. Here they are in fold units consisting of slatted-floored night arks with pens attached, moved every day to fresh ground, until the laying stage for pullets or fattening weight for cockerels, if they are not sold off the run. While this system was started primarily for the sole benefit of the health and wellbeing of the chickens, it is this moving of the units daily over all the land which is not used for breeding stock, which has trebled the stock-carrying capacity of the grass for grazing animals. Although the pens will not be on the same strip of land more than one day in the year, so valuable are poultry for reclamation, in scratching out moss, treading down rough broom grass, and leaving their rich manure behind, that on this poor light land it is now possible to carry a beast to the acre, grazing before and after the pens, where before three acres was insufficient, and had to be supplemented. Considered as a crop, allowing for the difference between their value at eight weeks and at five months, chickens are worth £100 an acre. Penning on stubble and fallow when available never fails to leave its mark in the following crops. Poultry manure shows for two years, nitrogenous effect the first year, phosphatic the second. It has been calculated that fold units leave behind the equivalent of four hundredweight of sulphate of ammonia, two hundredweight of superphosphate, and one and a half hundred-weight of potash, and in a far more valuable form, for every acre penned or two tons of food consumed. So highly do we value poultry for fertility purposes that we would consider their retention justified even if they did not leave a profit in their produce; though in actual fact, with good management, they seem absolutely foolproof in this respect, and we have never failed to show better figures year by year.

We have had our setbacks, mostly on the business side, but in the long run they have proved the old proverb 'When one door shuts another opens'. After a few years another enterprising poultry farmer undercut our price for table poultry and we lost our contract. A blow at the. time, but we soon found we could do as well selling our birds in Oxford, or retail to big houses in our own locality, which saved us the long run to London.

Early in the 1930's Mr. W. D. Evans founded the Kibworth Hatchery, and introduced to this country as a commercial proposition the sale of pure-bred sexed chickens, so that farmers would no longer have to rear the birds to eight or ten weeks old before the sex could be detected, or resort to sex-linkage, in which case the birds are useless for breeding. The subject and knowledge had been only a matter of scientific curiosity at the World Poultry Conference, when the methods used were described from the finding of two Japanese research workers at Edinburgh. But Mr. Evans, who incidentally introduced sex-linkage as a practical aspect in poultry farming some years previously, saw its great possibilities, and arranged with the Japanese Government to send an expert, thus being the first hatchery to offer pure-bred, sexed, day-old chickens.

Now, it has always been our rule, 'If you can't compete, co-operate'. So the day his first advertisement appeared we were offering regular supplies of Light Sussex hatching eggs. We with an output of a thousand chickens a week could not afford to employ a full-time sexer, and therefore could not compete with the latest novelty in poultry farming. Other big hatcheries soon took up the idea and it has become an established practice in the industry.

Selling our eggs, with the proviso that we could have back any chickens we required of our own breeding, proved a blessing in the long run, as hatching chickens is really a specialized branch of the business, and our change of policy enabled us to give more time to rearing. We could now raise twice the number of pullets in the spring, as space would not be taken up by surplus cockerels, while in the autumn we could rear double the number of cockerels for table and not have to run on the pullets with their slower growth which was always retarded by competition with the male birds.

Thus we started a long and happy association with the Kibworth Hatchery, and must now have sold them well over a million hatching eggs. Not only was this hatchery the first to introduce the sexing of chickens, but it maintained a strict health control scheme for the flocks from which its supplies were drawn. Nearly everyone must have heard how disease wrought havoc throughout the poultry industry in the middle 'thirties, when even Laying Tests, with the pick of the finest birds from the leading breeders in the country, had to report an average mortality in laying pullets of twenty and in some cases up to forty per cent. We ourselves used to look anxiously at our birds every day, half expecting some fell symptoms to appear, but our stock was spared, and looking back we are inclined to think that it was very largely due to the untiring efforts of the Poultry Pathological Research Laboratory in whose hands Mr. Evans had placed his scheme. We believe in giving credit where it is due, and in our opinion the disease control measures under the Ministry of Agriculture's Accredited Poultry Scheme are but a pale shadow of those rigidly enforced by the Kibworth Hatchery ten years ago.

So as the years went by we tended to concentrate on two main lines in the poultry industry, the sale of hatching eggs and reared pullets, selling some 120,000 of the former and 5,000 of the latter; and it is only since the outbreak of war that we have returned to selling day-old chicks and stock cockerels in addition to hatching eggs and fattened cockerels. These last, of course, are on a greatly reduced scale owing to the shortage of feeding-stuffs and low controlled price, deliberately fixed to discourage their production.

Wartime conditions brought many new problems to be faced, but we were much more happily placed than the majority of poultry farmers, and with our intensively farmed arable could at least produce our own grain; and we were determined to hold on to our stock, which represented to us a life's work, even if the Minister of Agriculture in conference with the farmers' leaders could throw the whole industry overboard in one short conference. A little thought would have shown them that the vast resources of British agriculture with proper organization could have maintained at least a, bird to the acre without any dependence on imported food and ensure at least one egg per person per week throughout the war.

Until the rationing system was introduced we had to maintain our birds by fair means and foul, as described in the chapter on wartime farming. By the autumn of 1941, when a new system of calculation was introduced, we had sufficient coupons to provide our 1,600 laying birds with half an ounce each per day, or one-eighth of their requirements.

As we were not permitted to retain our wheat, and our other corn, after allowing for cattle and pigs, was only sufficient for one and a half ounces per head daily, we had to find the equivalent of one and a half ounces of mash and half an ounce of grain to provide the usual four ounces for each bird. Being too far from a source of prepared swill, we had to look elsewhere for supplementary feeding-stuffs. Fresh vegetable products, carrots, mangolds, rape, kale, etc., can only be used in small quantities owing to their bulkiness. One to three pounds may be necessary, to replace two ounces of a cereal mash. As the capacity of a fowl's crop is four fluid ounces, normally half-filled twice a day, it is a physical impossibility for the bird to digest enough of these foods for maintenance and production.

Unrationed by-products of the milling and seed-cleaning industries are more useful, their only disadvantage being high fibre content. Supply is variable, but they store well and can be kept for twelve months in ratproof bins. Nearly all can be improved by grinding into a fine meal, and even such unpromising material as barley awns, oat husks, trefoil cob, and flax chives can be used.

Pig potatoes were our great standby, with their ratio of four to one compared with meal. Easily stored all winter, and ensiled in late spring for summer feeding, when steamed and mixed hot with other foods they go well with the more fibrous low-grade meals.

We were also fortunate in getting enough semi-solid buttermilk and whey, in barrels, to balance other foods which are deficient in protein. Malt bran and wheat germ were also available, and unrationed, to anyone who could take several tons.

The mash we used usually consisted of 40 per cent potatoes, 20 per cent wheatings, 10 per cent wheat germ or malt bran, and 20 per cent ground husks, weeds, acorns, horse chestnuts, dried grass, and similar substances in as great a variety as possible, to counteract any harmful properties any one may have had; the mixture was completed by the addition of 5 per cent fish or meat meal, and 5 per cent semi-solid whey.

This 'austerity' mash, compared with the average analysis of a good poultry meal, required two ounces to replace one and a half ounces. We overcame this drawback by feeding two feeds of mash per day, half at dawn and the rest at midday. With the grain feed as late as possible in the evening, too great a strain is not put upon the digestive capacity of the birds.

Production was maintained within ten per cent of our monthly average over a period of seventeen years and mortality and culls two per cent lower. Fertility averaged ninety-two per cent over the whole winter.

Then with the introduction of the Government's new accredited scheme and the granting of food based on the actual number of birds mated, the food position became much easier, and we were in a strong position to take advantage of the Domestic Poultry Keepers' Scheme, under which general farmers were to rear for the backyarders, which was introduced in the following spring. One year we were being told to scrap our birds, the next that they could render a very valuable service to the war effort.

The response to the Ministry's appeal for rearing the replacements was poor, and many missed the greatest opportunity that had presented itself to the hardest-hit section of the farming industry from the outbreak of war.

On our large stock of hens we were granted sufficient food to rear eight thousand, or twice the number we normally reared in pre-war days. This we could undertake without neglecting our trade in hatching eggs and disappointing old customers who might also wish to rear their share. We therefore reserved 900 eggs per week, to give us 600 chickens, 300 pullets, and 250 reared, to allow a fair margin for culling and mortality. If the early hatches did well, we should be able by the flexibility of the system in obtaining food to rear them, to stop a week or two before the end of the season.

Hatching and rearing was a simple matter of ordinary routine. Attention to detail had long since become almost automatic, so that the smallest water trough left unscrubbed or a brooder house window left open never escaped attention.

Office organization proved a bigger problem. We had not needed to advertise our stock for many years in selling to general farmers, but to obtain a different class of customer we inserted a series of small advertisements in a local city paper, resulting in a dozen telephone calls a day and twice that number of inquiries and orders by post; in all we handled just over a thousand orders, many hundreds had to be turned away, and only once did we fail to execute on the date promised.

The simple method devised was to have a sheet for each week, enter each order with the number of chickens required in a column, and as soon as the total of 250 was reached that week was closed. Forward bookings were entered on the appropriate sheet if a date was specified or at the earliest possible to ensure clearing the pullets at eight weeks old. In this way each week's delivery was together and the route could be planned.

Labour was the next problem. There were only four of us, instead of the usual five in past seasons. On top of the work of an ordinary mixed farm, we intended to rear twice as many pullets again as normal.

Greater efficiency and longer hours were the only solution. As will be explained in the chapter on labour, it had always been our rule that everyone should learn to do every job on the farm, and as a profit-sharing system was working no-one need fear that he would not get his reward.

To see incubators and brooders filled to capacity again, was to us, in itself, a great consolation. For two years we had seemed to be attending to the obsequies of a dying industry, just rearing sufficient to maintain the breeding stock.

Delivery, although requiring a lot of time and patience, was easily arranged. Shortage of crates and delay in returning empties by rail in the first few weeks necessitated delivery by road, and so, except for a few old customers we refused distant orders, and we were very glad we did when we found we could sell our entire output locally. All the birds being the same age at the time of delivery crates could be filled to capacity, and the small numbers counted out to each customer. Routes were arranged with the aid of a large-scale street map, which incidentally could only be obtained after satisfying the Chief Constable of our bona fides. How necessary a map is, will be appreciated from the fact that although the return journey to Oxford is only thirty miles, twenty deliveries within the city boundaries may double the distance covered in spite of careful planning.

The time and petrol used in making personal deliveries was well justified. Many newcomers to poultry-keeping required help and guidance to make the best of their birds. Generally speaking we were very impressed by the ingenuity and resource of the backyarder in devising a suitable pen and run; and as a snapper-up of unconsidered trifles with which to supplement his meagre rations he was unsurpassed. In only a few cases was human food being wasted, and it is hoped that these people realized the error of their ways when it was pointed out how easy it was to obtain the balancer meal and more economical to beg scraps from the neighbours. There were always a few who wanted to put eight-week-old pullets in with some old ducks, but found they would first have to find a suitable pen, as we would not allow them to spoil good pullets, even if they had paid for them; we hope they blessed us in the end. The odd customer who disclosed that he was only buying pullets to fatten for the table, learned that he could have cockerels cheaper, and good pullets would be found a home elsewhere. There is no doubt that the personal touch means a great deal to the domestic poultry-keeper, and any bread we cast upon the waters came back plum cake in their recommendations, so much so that we cleared every bird we reared, even when the Ministry suddenly changed their mind again and reduced the rations of the backyarder, although many poultry farmers were caught with birds on their hands owing to the inevitable cancellations.

The maximum prices suggested by the founders of the scheme were accepted without question, as it was generally realized that the old birds they replaced were worth nearly as much to kill. We calculated a margin of threepence per bird sold would be sufficient for replacements should any customer meet with misfortune in the first few weeks, for which, of course, we were not legally responsible, but with a fair margin in the price charged were prepared to meet. The actual cost of replacements was less than a farthing per bird sold, which is a great tribute to the care and attention which the backyarders must have given to their birds.

We had been specially warned against supplying poultry clubs, because if one member should be dissatisfied through any cause all the others would hear about it; but we did not find it so in practice and we prefer these bulk orders where there is an efficient poultry club secretary, who does quite a lot of our work for us, while saving members paying the full retail price, as we can make an allowance of sixpence a head in lots of a hundred or more.

In our opinion, it was a good sound scheme, well planned and thought out, carefully supervised by the county officials, and leaving a good margin of profit for efficient rearers. Without the scheme, the backyarders have been tempted into trying to rear chickens on good human food, and losing nine-tenths of them for one cause and another, and what they do rear cannot be compared with those raised under ideal conditions by poultry farmers who have specialized in the business for many years. This may sound like vested interests, but I do think that local education officials who have the melancholy job of advising the small poultry-keepers would agree that if chickens are to be reared it should be done by the specialists; and so now the Ministry have revived the scheme for 1944, and we again hope to rear our full share.

For the future prospects of poultry farming we are as optimistic as ever, but as there will be many ex-service men hoping to take up this branch a few words of warning based on our experience, which covers two wars and the period between, may not be out of place.

We do not believe that we could have made a success of poultry farming as a specialized business, but only in conjunction with a general farm. Fresh land and extensive methods are essential, if disease is to be avoided and production maintained. On this farm, the difference between one egg more per bird, each month, and one egg less per bird over the same period amounts to £800 per annum. A farthing per dozen more, or a farthing per dozen less, made a difference in normal times of £50, while five shillings a ton in feeding-stuffs made the same difference. So it will be seen how closely poised is the balance between profit and loss in poultry farming, and that a very high standard in management and efficiency is necessary. Full production, well-sold produce, and carefully bought feeding-stuffs, might leave a profit of £1,000 a year on a thousand birds; in fact we have made it, but this is only achieved by long experience, built up year by year as our stock increased, and unremitting care and attention to detail. Carelessly mixed poultry mash or running the birds short of water for a few days at one time of the year and putting the birds into a moult might easily cut the profit in half. An outbreak of disease might so undermine the health of the flock and reduce the numbers that there was no possibility of a living being obtained; so we do most strongly advise any prospective poultry farmers to learn their job thoroughly before investing any capital, and then start in a small way with sound stock, so that their experience is always equal to the stock they have to manage. But without some sort of general farming the man who starts in a small way to breed up his stock is not fully occupied, and nothing degenerates quicker than the mind and body that is not fully occupied, so that we do not recommend poultry farming as a branch in which to specialize, rather as the most valuable adjunct of all to general farming. And herein lies the easiest source of profit to the man who is master of his job.

While health and production have been maintained over a period of nearly thirty years, we have a system by which costs and production can be checked week by week and only on three occasions have our birds failed to pay their way. It should not, however, be thought that our poultry section is the model on which other farms should be based. A casual walk round by the Ministry of Agriculture's inspectors enables them to point out a dozen faults in our stock and management, which makes us wonder why they accept appointments in the Civil Service rising at the most to £950 per annum, when such omniscience should enable them to earn £10,000 a year as poultry farmers. But we have long since realized our limitations and perfection is something we may never attain. It suffices that we retain the confidence of our customers, for it is on them that our business depends, and not on any official approval. Though with the Accredited Scheme as such we find no fault, it is simply recognition of a reasonable standard which any reputable breeder would maintain for himself, and we would like to see it extended to all classes of stock.

# CHAPTER FOUR

## THE CATTLE

It is perhaps a little ambitious to found a herd of pedigree Jersey cattle on £50, but as we had no more capital available there was no alternative.

In the four years while the writer was learning farming he had had practical experience with six breeds, Jersey, Shorthorn, Friesian, Aberdeen Angus, Ayrshire, and Devon cattle. But the first was without doubt the most suitable for our requirements. The most economic producers of rich milk, in our opinion the most healthy and hardy of all dairy breeds, and with greater freedom from disease, Jerseys are ideally suited to the small farmer. As the demand for them was small—a country gentleman will buy one or two for household requirements—it was easier for us to obtain recognition as cattle breeders, for the grazier or fattener of other breeds will only buy in large even bunches, and a small farm, however suitable for stock rearing, cannot supply what they require. So the Jersey was our choice, and one we have never for a moment regretted.

We needed two cows to maintain a milk and butter supply all round the year for our own requirements, preferably an autumn and a spring calver, to bridge over the period in which each would be dry in. turn. The financial position put a limit of £25 a head on what we could buy. How easy it is to go to a sale and pick out the best animal, but how difficult to decide which is the best that can be obtained for a certain sum. We went to sale after sale without buying. Mother used to laugh at us, and say, 'When you go to a sale you never buy anything'. However, we finally got what we wanted.

I bid to our limit for cow after cow, then one was knocked down to me. The keen-eyed auctioneer had missed nothing, for he smiled across the ring to me. 'You deserve to have that one,' he said. 'You have been trying hard for a long time.'

A little later we were again lucky and bought another first-class animal for twenty-four guineas. Once more came the quick smile from the representative of the famous firm who have since sold many, thousands of pounds' worth of cattle for us, but on that day he seemed as pleased to be selling me good animals at a price we could afford, as to be piling up the death duties for the famous millionaire for whom he was selling. Though I may say I have never known these people to fail in their duty to a vendor, while holding the balance fairly between buyer and seller, in striking contrast to some auctioneers who reduce their profession to the level of the cheapjack bellowing in the marketplace.

*LAMBS CREEP FORWARD FOR A LITTLE CONCENTRATED FOOD*

*A PUPPY LEARNS HIS TRADE*

*TEN DAYS OLD*

*INTENSIVE PRODUCTION: SHE STANDS GUARD OVER THE*
*BACON RATION FOR NINETY PEOPLE FOR A WHOLE YEAR*

Josephine and Evelyn; how we loved those two cows, and how well they
served us. They cost perhaps twice the price of ordinary commercial cattle

at that time, but probably only half their pedigree value, for one was in poor condition through heavy milking and the other a long way off calving; and being old, did not catch the buyer's eye. But to us they were invaluable, for they were the foundation stock of the 'Enstone' herd ('Enstone' being the prefix which every animal born here bears before its name).

How anxious we were that they should travel home safely, for to us they were irreplaceable! How carefully we unloaded them, cleaned and fed them! Neither had given more than 700 gallons, but with modern rationing and three-times-a-day milking, both went over the 1,000-gallon mark, besides breeding us several good waives. Josephine bred and milked till she was fourteen years of age, and Evelyn until she was eleven. From their milk we sold an average of twenty pounds of butter a week, which in those days was sufficient to pay the tradesmen's bills, and we had the separated milk on which to rear calves, pigs, and chickens. For our care and trouble they repaid us many times; so quiet were they that we could take a bucket and stool out into the field and milk them anywhere, so docile that we could tether them on arable crops of clover, vetches, or kale.

Through the kindness of a great landowner we were able to put our cows to a first-class bull, but how slowly a herd grows for those who rear their own stock. It took ten years to increase to ten animals. Our system, then as now, was to sell calved heifers, retaining their calves to maintain the herd. However, in our early years we bought and fattened for veal a few beef-type calves of other breeds as a sideline. These calves were always carefully isolated from the Jerseys for fear of disease, for we were determined to have a disease-free herd right from the start. These beef calves would take up to three gallons of Jersey milk a day and increase their weight by as many pounds, and with veal at one shilling and threepence a pound, we could sell them at ten weeks old for a ten-pound note. At that time many farmers were bemoaning their fate in having to sell surplus milk to the milk factories at four-pence to sixpence a gallon. We were cashing ours through calves at one shilling and threepence. Also in making ten pounds on a calf at ten weeks, we were realizing as much as the same animal would be worth run on to fifteen months old and sold for a store beast for fattening; and even then would lose money when sold for beef at fifty shillings a hundredweight.

However difficult the times were, we always believed in looking for an opportunity in every difficulty, and never for the difficulty in any opportunity. If we had to wait for our Jerseys to grow into a herd, we could snatch a profit from the depressed beef industry, though it was always our rule never to buy at a price that would lose the vendor money. If we could not bid a fair price for a calf, we would not buy it at all. The best veal is a luxury trade, the calves must have a lot of milk, no cake or hay, and be sold just before their horns come through. Leave them a week too long, and their value is reduced by fifty per cent. On one occasion we were shut up by foot-and-mouth disease in the district and had to run some calves on; we fed them well throughout, yet as baby beef at six or seven hundredweight they realized no more than they would have been worth for veal a year previously.

At about that time we were approached by a well-known breeder of Jerseys, who had heard that we were very successful calf rearers, who offered us a fair price to rear all his calves and return them as heifers. The reason was that he was experiencing very heavy mortality, nine out of ten, for no cause which could be determined. We accepted, took most stringent precautions in isolating them from our own stock, and reared ninety-nine per cent. Any heifers that were not required back in the herd we could sell, and share the profit over and above the agreed price for rearing. The whole system proved a very happy and successful venture on both sides.

We were then able to concentrate our whole attention on rearing pedigree Jerseys, and did not fatten any more calves of other breeds for veal.

In a few years we were able to buy good bulls of our own, which was justified by our herd of forty to fifty animals, and we no longer had to take calves for rearing, though we continued to buy from the herd for which we had had the pleasure of rearing many fine and famous animals.

With our herd established, our system was, and is, to rear twenty to twenty-four heifer calves each year, selling them calved at two years old, retaining their calves to maintain the herd. Four of these first-calf heifers are kept back to rear the next lot of calves, and sold when calving for the second time. In this way we never have an old cow to dispose of. If extra calves are required to maintain the herd these are usually obtained from producer-retailers of milk who buy our heifers but do not wish to rear calves, although they sometimes reserve the option of buying back, when reared, the calves they sell us. This we are quite happy to do, for our job is calf rearing, the farm being unsuited by soil and location for milk production.

In recent years the importance of animal health has been stressed by many eminent authorities. The writer realized it while still a pupil, when seeing the heavy toll which disease levied from all classes of stock. It is our proud boast that every animal sold has been offered always subject to any health test the customer may desire. Starting with tuberculin-tested animals in the first place, in the days of the old thermometer test, we had our first intradermal tests made in 1927, continued with it, and in due course were accepted for the Attested Herds Scheme under the Ministry of Agriculture, when it was introduced. We cannot pay too high tribute to the veterinary inspectors administering this service to the industry, whereby farmers can buy sound stock under what is tantamount to a guarantee that it is free from tuberculosis. We would like to see it extended to cover contagious abortion, mastitis, and trichomonas.

Bloodtesting for abortion here has long been a matter of routine. No reactors having been found, we do not intend to take up the new immunization method introduced by the Ministry in 1943. Mastitis and trichomonas we believe to be the result of mismanagement and unlikely to occur with efficient supervision, unless they have become endemic in a district; and we are fortunate in that we farm in arable country and not in a dairying area, where the strictest precautions are necessary to avoid infection. It has been our experience, and is backed by the opinion of our local veterinary surgeon, that a system which can keep stock clear from the more dreaded diseases ensures freedom from the lesser complaints to which cattle are liable. In the twenty years in which nearly five hundred Jerseys have passed through our hands we have only lost four calves and one cow, the latter by a post-parturition infection, which indicates that Jerseys are not so delicate as some would contend.

For economic rearing and production the Jersey must be unsurpassed. From birth to calving a heifer requires fifty gallons of milk, a ton of hay, two ton of roots, and six hundredweight of concentrated food, or its equivalent in silage, and half an acre of good grazing. This is less than half the food required by any of the heavier breeds, allowing for the fact that they go another nine months before calving. Fifty gallons of milk may seem a generous start to give a calf, in view of the campaign by the the Milk Marketing Board and the Ministry to persuade farmers to rear on substitutes, but we believe this to be the most short-sighted policy ever evolved by the bureaucratic mind. How much wiser it would have been to ensure the future wellbeing of the bovine population by advising better feeding of calves on milk produced by more efficient management. One good calf is required to replace four cows, and there can hardly be a cow in the country whose yield could not be increased by ten gallons per annum by careful management, and this in addition to the normal allowance would be sufficient to ensure the health and wellbeing of wellbred calves, on whom the future prosperity of the industry, and the health of the nation, depend. I once had the pleasure of studying some records kept by a farmer in Northern Ireland over many years, which showed conclusively that for production and longevity the cows which had been reared on two outlying small farms, with ample milk available, had far surpassed those reared under the farmer's own careful supervision on calf meal, where he was in a position to retail every pint of milk produced, and had therefore been misled into believing that calf rearing with meal was more economical. It is useful in changing over from milk, and we use it, but we still believe that the cow will return a hundredfold the milk she received as a calf, or will withhold in proportion that part of her birthright of which she has been robbed, through the reduction in health and vitality resulting from it. It is not what a calf looks like at six months old that matters, it is how much milk she will be giving at six years old or later.

For milk production a well-bred, well-reared Jersey would be hard to beat. We have had them giving forty pounds of milk per day on thirty pounds of food, and their own weight in milk in seventeen days, their own weight in butter in a year. So highly bred are they for milk production that some will come into milk three months before having their first calf, at twenty-one months old, without apparently suffering in health; while we have one at the moment which has been milking for four years without going dry, has had two sets of twins in that time, and is still in full production. We do not normally keep a cow for so long a period under our system, but on one occasion when she would have been sold, she knocked a horn off on the day she was due to go to the sale, and on another the sale was cancelled by foot-and-mouth disease outbreaks.

While these are perhaps abnormal examples, the yield of milk from the average heifer, now in the ninth generation, has been increased by fifty per cent by careful breeding and modern methods of feeding. Sterility, a common fault in highly bred cattle in bygone years, has disappeared thanks to the resources of science and the more natural methods of feeding under war conditions.

The buildings on the farm, of which more will be said later, have been designed and built to accommodate the herd. Having made our plan and calculated the numbers the farm should be able to support, the. building programme was dependent on increase and profits the stock could earn, owing to our rule that each department can only extend out of profits; and while it may involve a little overcrowding in the early days, if the cattle have to pay for their buildings before they can occupy them, it is very comfortable and convenient when the herd is established.

Concrete, wood, and galvanized iron are cheap and efficient; and can be pleasing to the eye, if painted green outside, white inside, with creosoted posts, rails, and gates. The criticism has been made by visitors to the farm that there is nothing permanent with these materials, compared with brick, slate, and stone. But why should we inflict our architectural ideals on someone who may farm the land in a hundred years' time? There has been far too much of this in agriculture, especially if it saddles the land with a heavy charge, or alternatively is never paid for. We have the satisfaction also that if our buildings were bombed or burnt they have already paid for themselves many times over. Also with our own construction in these cheap materials, we can have substantial covered yards, while the farm with its old-fashioned buildings wastes in the open yards half the value of the manure produced, and sometimes a great deal more in a wet winter. How pleasant it is to have properly arranged feeding passages, hygienic calf pens, tubular cow stalls, safe bull boxes, in which the animals can be caught and handled without entering, and convenient water taps and hoses exactly where they are needed.

From the financial point of view the herd has never failed to bring in a steady income, and increased in value from £50 to perhaps £3,000 at current values. Who can say that pedigree breeding is only a hobby for the wealthy man? True, it has taken us half a lifetime to establish, but how worth while! Every heifer sold with her name, registered number, and prefix, is a free advertisement wherever she goes. The rearer of commercial cattle may make a profit or loss according to the state of the market when each bunch is sold, but then has to start all over again, while the pedigree breeder builds a name and a reputation if his stock is worthy of the breed he believes in. How thrilled the writer was, when over a thousand miles from home, and in reply to a question, he said that he came from a little place of no importance called Enstone. 'Enstone?' said the stranger. 'Is not that the place where the lovely cows come from?' This person had been a governess in a ducal household in England, and having been told the name of two cows, 'Enstone Golden Mist', and 'Enstone Early Dawn', had remembered the name, without suspecting for one instant that she was talking to the man who had named them. Such is fame, when your stock is better known than you are yourself, and for a farmer it is as it should be.

Over the years we have made many good friends in the Jersey world. How pleasant it is to show people our stock, or to visit and inspect their herds. How we enjoy helping a new breeder to select his stock, and somehow the financial aspect seldom seems to influence us. There is obviously a good margin of profit in this type of stock, and it is very nice to show better figures when the books are balanced at the end of the year, but in the long run it seems certain that they who covet profit least, profit most.

I remember once a wealthy farmer coming to try to buy a milking heifer. I showed him two; one at thirty guineas and the other at forty, which was then the current price and good honest value at the money. He found fault with both, where faults did not exist, but did not mind which he had providing he could beat down the price a couple of pounds; not that he could value either within a fiver, but he was going on the old-fashioned farming principle, that you cannot do business without a haggle. Unfortunately for him, it is our strict rule that we never drive a bargain, but offer or ask the true value; and so he did not buy.

The next day, a young couple came to look at the same heifers. The man knew his job. He studied the animals very carefully and then went aside and talked to his wife. Then he came back and said, 'I think they are worth every penny you are asking, and would have made it at the pedigree sale at Reading last month, but we are only just starting and have not the money available to buy the best, so we will have the thirty-guinea one.' 'No,' said I, 'you will not, nor will I insult you with an offer of deferred terms or hire purchase. You will have the forty-guinea heifer for thirty, and may she start a herd which will serve you as well as ours has done for us.'

We do not of course always do this; normally our price is the same to prince or pauper, but it is nice to help an honest man when you find him. Whether this action was quixotic, bread cast upon the waters, or ground bait, I will not pretend to say, but it has come back a hundredfold in the recommendations these people have given us, and the stock we have sold them since.

Whether pedigree stock breeding can be considered a democratic institution we do not know, but it is very agreeable to simple farmers like ourselves to be treated as equals by real gentlefolk, whose interests coincide with ours in breeding finer and better stock. The little jumped-up profiteer comes sailing in, 'Ah! I am Mr. Blank, and this is my daughter Miss Blank!' The bearer of a famous name, with a dozen titles says, 'I'm A——, this is my son Henry'. The latter of course may be a marquis or an earl, but at the moment his father has brought him along to learn something about choosing a good cow, very useful knowledge for one who will have a real interest in his tenantry. We know our place, we share our knowledge freely, and feel that it is a privilege to be of service. On the other hand our Mr. Blank knows so much that we do not need to tell him anything, and he soon sums us up as a couple of ignorant country louts; we are sorely tempted to ask him inflated prices for anything he fancies, but somehow resist it, and belittle our own animals so that he does not buy.

The other type of buyer we do not like is the one who pokes the cattle about with a stick. So deeply ingrained is the stick complex in some people that they always associate it with cattle, not realizing that it is not even necessary to shout at them. All our animals can be handled in the field, and led out of the herd. A cow that will lead quietly, probably yields an extra hundred gallons a year, compared with one which has to be driven, for she associates being led with kindly human beings, just as a dog likes being taken for a walk; and there can be no doubt that the psychological aspect has a very great effect on a cow's output of milk. In any case, it is good business to have quiet stock. The heifer which comes out of the herd when called, and in the case of a lady licks her hand, is more than half sold.

The extreme case is illustrated by the man who called when I was carrying a bundle of pitchforks out to the hay field. 'What do you do for forks in wartime?' he asked. 'The same as we have done for twenty years,' I replied in surprise, 'for these are the same.' 'Oh,' said he 'I have to have a new set every year. My fellows break them on the cattle each winter.' This, I felt, was on a par with the man who told me that the great advantage of the milking machine was that one did not get complaints of cigarette ash in the milk!

However, it takes all sorts to make a world and we noted recently that a well-known writer suggested that a charge of shot (naturally from a safe vantage point) was an excellent quietener for a savage bull. Personally, as with cows, we prefer the psychological approach, and on one occasion had the experience of comparing the two methods.

A gentleman had bought a bull at the pedigree sale at Reading, on our recommendation, from a herd in charge of a man with an excellent reputation, who handles cattle on the same lines as ourselves, and never has any trouble with them; although Jersey bulls are reputed to be notoriously vicious, in striking contrast to the docility of the cows. One morning we received an urgent telephone call to say that the bull in question had broken loose and was terrifying the whole village. Could we do something about it?

My brother and I motored over, carefully laying our plans on the way. On arrival we found the bull cornered in a field, into which he had been driven by men with shotguns and pitchforks. The owner was prepared to have the bull destroyed in spite of the high price he had paid for it, for it roared furiously and tore up the ground, as the men menaced it with their guns, one or two shots having been fired at it. To this plan we could not agree.

First we insisted that the guns should be unloaded, and everyone was to get out of the field. This they did, but peered over the high wall to see what we would do. Then we strolled into the field, about ten yards apart, taking no notice of the bull, who again roared at us as only Jerseys can (for it must be remembered that they have a common origin with the Spanish fighting bulls, both belonging to the order *Bos sondaicus*, or the ancient Iberian cattle of the Mediterranean).

A dozen yards from the gate I saw a piece of old binder twine on the ground and stopped to pick it up. My brother took his cue, stopped also, but watched the bull closely. I turned to the watching men. 'This is very dangerous stuff to leave lying about,' I said severely, 'if the bull ate it he might get a stoppage.' This raised a laugh. My brother signed that the bull had approached, and was looking curiously at me. I slowly turned, and addressing the bull, I said kindly, 'Well, old fellow, do you find liberty overrated?' He drew back a pace, eyeing me suspiciously, snorted, and came a little closer; this was the sort of human being he was used to, but what about the shouts, bangs, and shooting pains in his hindquarters? We continued to look at each other for a few minutes. Then he came a little closer. My brother sat down on the ground, the bull turned his head to look at him in surprise, and I caught his head chain and nose ring, and he submitted himself to be led quietly out of the field.

Any risks we ran were negligible, for at no time were we more than a few yards from safety; had the bull offered to attack either, the other would have diverted his attention sufficiently for us both to vault over the wall. We had no intention of venturing far into the field; had the string not been available, I should have put my watch right with elaborate unconcern, or inquired if the audience had heard the story about the actress and the bishop. Not that I had one to tell, but it was essential that the bull should approach. We relied chiefly on the fact that the bull would be tired, cold, stiff, hungry, and longing for someone to lead him into a nice warm loosebox for food and rest, but nevertheless would defend himself, as nature intended he should, if needs must.

Had this been an animal which had been running loose, and perhaps teased by village lads throwing stones at it, or frustrated by having cows taken from him, then our method would have been suicidal; as it was we could have our little bit of fun with a show of bravado.

In handling bulls the great thing is to have a regular routine, take them for a walk every day, brush and comb them, so that they are used to being handled. Never frustrate them. When taking them into a loosebox after exercise or a service see that there is a feed waiting for them, so that they associate going in with being fed and not with being shut up. We do not use a bull stick, simply a divided chain, one spring hook going on the head chain, another on the nose ring, so that the bull is led on the head chain, and is only pulling on his nose ring if he puts his head down. The bull is always led on the right of the man. If he comes towards him, the chain is rattled and the bull shuts his eyes and draws back. In turning the bull goes round the man, never the man round the bull. In this way the man never need be crushed against a wall or gate, at the worst he will only be tossed, and speaking from experience that is no worse than being thrown from a horse. To struggle about with a bull on the end of a bull stick is to have only an illusion of safety, because a stick may break and many accidents have been caused in this way. A bull that has been brought up on chain leading should never be led with a stick, or vice versa. However quiet a bull is, and the vast majority are if properly handled, one must never forget for an instant that they are like a kettle of boiling water, or a loaded gun, and only carelessness will cause an accident in ninety-nine cases out of a hundred. The hundredth is when, say, a low-flying aeroplane roars over the roof tops and the bull shows his resentment in the only way he knows; his proud masculine spirit will not let him flee and so he turns on the man before the chain can be rattled, the sign he has learned from youth to obey, and slings him out of the way. I have been thrown on to a roof, slid off again, caught the bull, and led him into his box, without suffering more than a few bruises, and without letting the bull realize that his strength is vastly superior to that of the puny creatures who control his destiny, for it is a thousand pities to have to sell a bull because he is getting mischievous, when he could still leave some very valuable stock. One of the finest bulls we ever owned, was bought at killing price because he had become morose and savage, through being shut up in a dark loosebox, and only let out for service, and poked back in with pitchforks; yet he bred us twenty-two fine heifers. For three days he hurled himself at the steel-lined and reinforced box prepared for him, refused to drink for a week, yet became quiet and docile with humane and patient treatment.

The one thing we have learned about bulls, if nothing else, is that a pennyworth of knowledge is worth a shillingsworth of courage, or a pound's worth of brute force. In common with boars and stallions, they cannot be trusted, and it is best never to give them an opportunity to do harm, by carefully planned routine and management. Possibly the most important thing of all is never to be afraid of them; the man who is, is probably much better at some other job. Personally I would sooner tackle two angry bulls than one angry man—while the ink on my pen congeals at the mere thought of an angry woman!

There can be no doubt that there is immense scope in this country for breeding better and healthier cattle, but any improvement must come from the farmers themselves, and not from any form of Government control; though the Ministry could render valuable service to the industry in an educational sphere.

Bull licensing by the Ministry livestock inspectors is foredoomed to failure, and only antagonizes those it is intended to help. It breaks down before the human factor that no good judge of one breed could afford to take a position at £6400—£650 a year, and that man seldom judges by appearances alone. The inspectors are expected to judge any one of twenty-four different breeds, to say nothing of crosses, which may be put before them. One of the greatest breeders of all time, Thomas Bates of Kirklevington, asserted over a hundred years ago, 'a hundred men may be found to make a prime minister to one fit to judge the real merits of a Shorthorn'. Yet the Ministry had no difficulty in finding inspectors, whose main qualification must have been a willingness to accept a nominal salary.

In our experience the inspectors look for beef, even in a Jersey, and will pass without question a fat, nice-coloured bull. We once had one which we desired to keep back in growth and fed it accordingly. It so happened that both the inspector and a higher official called to license it. They were dubious about passing it, and in the course of the argument which took place I asked them to point out his chief fault. After some hesitation, they pointed to the bull's high-set tail, a characteristic of that particular strain associated with high milk yield, but not passed on to the heifers; the feature would not have been conspicuous had the bull been fat. However, they did finally license it, and it bred us £2,000 worth of stock in a year, and sold for three figures to a breeder, whose worst enemies could not accuse him of wasting money on bulls, and he used it successfully.

After twenty years neither my brother nor I can choose a Jersey bull on appearances. If we can see the herd from which it comes, the cow from which it came, the bull that bred it, and if possible his stock, make a careful study of milk records and pedigrees, and a close examination of the animal offered, we are able to say 'This may do'. Yet the Ministry's inspectors might condemn our choice on a casual inspection.

I have no grudge against them; they have to live, they are probably kind husbands and good fathers; but I do think they could be better employed in enforcing the Warble Fly Order, and rendering a far greater service to the industry in so doing.

In our experience the return from rearing good stock is quite equal to that obtained by the milk producer, and is well suited for small isolated farms, where the collection of small quantities of milk involves the use of transport and petrol which cannot be justified under wartime conditions, and is economically unsound at other times. Also the individual attention which can be given produces a far better animal for milk production than is possible on the larger dairy farms, where young stock is apt to be regarded as of secondary importance and is treated accordingly.

At the same time we do not recommend anyone to rush to the pedigree sales and buy an expensive bunch of calves in the belief that they have only got to rear them, put them to a good bull, and sell them for fancy prices. It is not nearly so simple as that! Not only must you have good stock to offer, but you must have a reputation for good stock. While we have never realized the record-breaking prices of some individual animals, few Jersey breeders could claim to have sold so many animals of a similar class, for so high an average, over so long a period. There are two reasons for this: the first is that most breeders do not sell twenty good heifers a year—they keep them in the herd; the other is that we have always specialized in this class of business and at every sale there are usually half a dozen people who are familiar with our stock and will bid it up to their limit—which enables us to maintain a good average.

We claim that our cattle are as good as or a little better than they look. Running out winter and summer, we never rug them up for the sale, which appeals to the buyers who are looking for health and constitution. Also we only spend six hours per head preparing them for sale, unlike the six weeks of some breeders, so that the customers are not disappointed when the spit and polish wears off. It is rather interesting that every 'Enstone' animal, re-entered at a later date at a collective sale, has realized more than we were originally paid for it, which is as it should be. Nothing is too good for the man who will pay us a good price for a good animal, for by so doing he is helping us to breed better and finer stock. Mutual service for the reasonable profit of both is far better than any satisfaction derived from the driving of a hard bargain, or the palming off by some sharp practice of something which is useless to both buyer and seller. In the same way, if by neglect or mismanagement a fine animal is ruined, long after we have sold it, we regret the loss almost as keenly as the man who owns it, and for that reason we are always willing to help or advise if we can see our way to do so.

Our belief that there is ample scope and opportunity for cattle breeding in this country will be borne out by a casual visit to any market. Even the Minister of Agriculture recently commented on the poor stock he has seen in travelling about the country, presumably bred under the Licensing of Bulls Act of 1931, which again supports our opinion that any improvement must come from the farmers themselves, In our experience the breeding of good stock can make the poorest and most isolated farm reasonably profitable over a period of years. A good knowledge of one's animals is more valuable than the longest purse; we cannot recall a single instance of anyone who has begun cattle breeding by the mating together of expensive animals ever making a name in the industry, while we could name a dozen who have bred a fine herd from a few carefully selected heifers mated to a first-class bull. For this reason, as shown by our experience, pedigree breeding can be within the ambition of almost any farmer. All he requires is time and patience, and he can laugh at the competition of anyone who rushes into the business in ten or twenty years' time, however much money the newcomer may have to throw about. Against which, the producer-retailer of milk always has to fear competition and the huge combines, which can spoil a life's work spent in building up a business.

But what if foot-and-mouth disease wipes out the pedigree herd? That is a tragedy all must be prepared to face. Personally I should start all over again, preferably with a few good animals rather than try to set up a whole herd on the compensation received; for good stock, acclimatized to the farm, can only be bred on the place.

As with all stock there is a limit to the size of the herd which can be profitably maintained. If we had twice the land and accommodation available we should not double our Jerseys, but would rather build up a beef herd, as in all our experience and travels we have never seen a dual-purpose breed. We remember once a newcomer to farming asking our advice. He had taken over a herd of forty Shorthorns, and was finding them unprofitable. Should he change to Jerseys? We carefully studied his farm, stock, accommodation, situation, and labour available. We recommended that he should keep thirty good Jerseys for milk, and twenty Aberdeen Angus cows rearing their own calves. With the same land, food, and labour as before, he produced more and better milk, more and better beef, turning a loss of £150 into a profit of £ 300 a year, which has steadily increased ever since. Specialization pays, and although many capable farmers will disagree, you cannot get milk and beef from the same animal. If she milks long enough to justify herself, she will be poor-quality beef, and if she is going to be beef the great carcase has to be kept up and maintained during her milking life; and even a bullock is the poorest converter of food into edible meat of all the farm animals, and very few cows can possibly pay to fatten for beef.

We have always looked ahead and planned for the future. We have every faith in Jerseys for the years to come. We believe that the day will come when milk will be sold on quality and Jerseys with their high butter-fat content of five per cent and over will become the premier milking breed of England, as they are of New Zealand. Meanwhile for economic production they are unsurpassed, are still comparatively cheap compared with other breeds, so we can recommend them with every confidence to any farmer who is prepared to give them the individual attention they require and deserve, and for which they repay a hundredfold.

# CHAPTER FIVE

## THE SHEEP

If it should be contended that farming to us is just a money-making racket, our sheep disprove it. True, they have always paid their way, but there has never been much money in sheep, because a sheep's worst enemy is another sheep, and any flock will die down to the capacity of the farm if overstocked; so that it is impossible to maintain the intensity of production necessary to make good profits. Whether farming is a trade or a profession depends on the attitude of mind; if the definition rests on the importance of the financial return or the satisfaction derived from doing a job well, then I think we can contend that we are shepherds by profession, however anxious we may be to turn an honest shilling in other departments of our business. If farming is a business, shepherding is still a craft. I have met some to whom it is a religion, but to us it has always been a source of pleasure which cannot be judged by the balance-sheet.

That sheep are necessary to make the best use of light land there can be little doubt, both in manuring and consolidating the land, clearing up kale after cattle (for the second growth even in a mild winter would seldom justify cutting again), and in grazing the stubble, catch crops, and clover leys, and so maintaining a properly balanced system of agriculture by returning to the land that which has been taken from it.

With fifty pounds to invest, it took us five years to build our flock, which was also the time necessary to get the land into sufficiently good heart to maintain them. We started with a dozen ewes and a borrowed ram; we did well the first year in rearing nineteen good lambs, but were compelled to sell them as ewes and lambs in the early summer, as we needed the money to tide us over harvest, and so could not make the best of them. In the next autumn, having sold our corn, we could again spend £50, plus the profit we had made on the first flock, which had been lent to the general farming account. We bought eighteen pedigree Hampshire Downs—who could say we were not ambitious?—and they certainly did well for we sold them for £7 each in the following spring, because of the financial difficulty of trying to farm well without capital, and heavy expenditure on good seeds and generous artificial manuring again ran us short in the months before harvest. But this proved a blessing in disguise, for the departure from the Gold Standard by the Government caused a slump in sheep which more than offset any further profit we might have made on them.

At the lower level of prices by the autumn of 1925, we saw that the profitable day of the Downland sheep, with their big joints which were no longer in demand, and their enormous appetites which had to be satisfied, was now over. We should have to adjust our methods to the times and make full use of the lighter, so-called grass sheep, which the public taste demanded, and which could still play their part in maintaining fertility. This was our chief concern, for all around us we could already see the declining returns on the farms which had given up their sheep in the Great War, in the mistaken belief that sheep did not pay, little realizing that the dividends of fertility which might have been stored up in the prosperous years, had the flocks been maintained at all costs, would have carried them over the lean times which were bound to come.

Border Leicester sheep were our choice. We would dearly have liked to have had a pedigree flock, but unfortunately there was no demand for the pure-bred rams in this district, as it pays better to put a Down breed ram with the north-country ewes, and so combine the good qualities of both; and so our plans had to be adjusted to circumstances.

So we bought twenty 'theaves', as they are called in this district, 'shearlings' in the north, actually maiden ewes about fifteen months old and ready to go to the ram in the autumn and have their lambs at two years old. These we put to a Suffolk ram, and it is interesting to note that recent experiments by one of the university farms have shown this to be the most economical cross of all, thereby confirming our choice.

The thrifty, active, deep-milking ewes, with the high fecundity of the Suffolk ram, give a high percentage of strong, healthy lambs, which can be sold fat when twelve to fourteen weeks old, weighing five stone, or run on to mutton weights in the autumn. The ewe lambs are also in demand; for some farmers put these again to a Down ram, for they too combine the good qualities of their mothers with the ability to lay on flesh inherited from their father.

The following year we again increased our flock, for this time the sheep did not have to help the farm, so that the old stock could be retained and fresh ewes bought with the profits on the lambs.

By the fifth year we had forty ewes, which means 100 to 120 sheep when the lambs are with them, and this represents the full capacity of the farm. On two occasions we have exceeded that number, and each time the number of lambs reared and gross value of sheep sold was lower than the average, proving that there is a limit to stocking with sheep quite apart from the question of feeding them; for these were bountiful years in which we had ample food available, so that limiting factor did not apply.

Once the flock was established it became our practice to buy twenty fresh ewes each year, selling those which had been bought two years previously. In this way we never have old, broken-mouthed sheep to dispose of, but sell them at their very best after two crops of lambs, or as the farmers say, 'four-tooths'. For a sheep's age is recorded by its teeth. For all practical purposes it has two large teeth in the centre of the lower jaw at two years old, four at three, six at four, and eight at five, when all the small milk teeth have been replaced, after which the animal rapidly depreciates in value as it loses its teeth, and is said to be broken-mouthed.

It has always been our practice to reinvest the money obtained from our 'four-tooths' in young maiden ewes from Scotland. If trade is bad, then our sheep do not realize quite so much money, the stock we buy is also cheaper in proportion, and the same applies on a rising market in a boom period. In this way the capital invested is always safeguarded; for all practical purposes we simply exchange our old ewes for young, the difference being that the former are proved breeders, while we give our skill and knowledge to making the best of the latter, having their two crops of lambs for the food and trouble, while the older ewes are better for the farmer who cannot give the same care and attention, or whose land is unsuitable for sheep with their life ahead of them. For to put young ewes on bad sheep land is to ruin them for life and heavy losses will be experienced, while sheep on these healthy Cotswolds will thrive with good management at any age.

In our early days it was our practice to spend on concentrated food for the ewes and lambs the sum realized by the sale of wool in the previous year. If it sold well they had more cake, which gave us more wool and better lambs. That which came from the sheep was returned to them, while the manurial residues helped to build up the fertility of the farm. Many farmers grudged their sheep this concentrated food, for which no stock pays better, and they were always the first to moan that there was no money in sheep; for in bad times it is always the average and the poor stock which is a drug on the market, the best have already been sold, the early lamb being like the 'early bird' in this respect. With the building up of fertility, and the winter green pastures due to heavy stocking with poultry, the time came when our sheep no longer required the concentrates; for this we reaped our reward when under war conditions no cake was available for sheep, as we could still produce fat lambs on roots, hay, and silage.

Up to the war we made it our rule to sell the lambs as soon as they would realize £2 each. With good lambs this might be at twelve weeks old, with poor triplets it might be necessary to run them on for six months. In this way the land was cleared of the best, giving the second grade more and better food to bring them on, while ensuring us our profit, for the cost price of producing a lamb on this farm, over a period of eighteen years, was £1. Only twice in the whole period were we compelled to accept less than our price, and this was more than offset by extra good prices in the good years when they made more than we expected, because the price had risen from the previous year by the time we had them ready for market. It is always a sound rule in farming to take a profit as soon as it is possible, rather than gamble on an increase in price or weight, the latter usually being cancelled out by bigger offerings in the markets and the consequent lower price, while the farmer gets nothing for the extra food consumed. Under war conditions we keep our lambs to make £3 apiece, and with a high percentage of twins and triplets we still find them remunerative, for costs have not increased to any great extent as they live entirely on the products of the land, now that nothing can be bought for them.

One of the most firmly held superstitions amongst shepherds and farmers is that one should not count the lambs, presumably for fear that there should be less next time they are counted, and in this way they do not know their losses. Personally, we consider it very unlucky *not* to count the lambs as they are dropped, as we want to know the losses incurred by weather or season, so that they can be remedied another year by better management or more careful planning. For there can be no doubt that there is very great scope for better shepherding, and lamb-recording in the flock can be as valuable as milk-recording in the cowshed.

Much useful information can also be obtained over the years by recording the ewes as they go to the ram, for the percentage of lambs dropped is more dependent on the management of the ewes during the breeding season than on any other factor. While it may vary considerably from farm to farm, it has been our experience here that if the flock is moved on to fresh clover ley on the day that the first ewe takes the ram, eighty per cent of the ewes that conceive seven days later, and after, will yield twins or triplets. This practice of 'flushing' the ewes is well known, but without careful recording of the ewes as they take the rams valuable data are lost. Constant observation of the flock is not necessary. The ram's chest is dressed with a mixture of powdered paint and oil, so that in serving the ewes he leaves all the evidence required. If in counting the flock each night the number marked is recorded, one knows that they will lamb in approximately the same order. This also gives the advantage that one knows what to prepare for in the lambing season as the sequence will be the same.

If the lambs are booked in day by day one also knows how many there should be at any given time, for many a lamb is lost by creeping through a hedge and being unable to find its way back. Or if a lamb should die without being noticed for want of being picked up and buried it may teach a fox or stray dog that here is an easy meal; and the animal will not be contented with dead lambs for long, but will plague the farmer until it is shot out of the way, causing many hours of waiting about which could be far better spent in a busy season.

Our percentage of lambs dropped and reared has steadily risen over the years, and the last two seasons we have managed to rear two lambs for every ewe that went to the ram. This may be a special wartime effort on the part of our flock to co-operate in our endeavours, but we hope not, for we shall need them in the hard and difficult years to come; for in all the changing fortunes of British agriculture we believe that the 'golden hoof' means more to the farmer and the land than any 'Gold Standard' which may be devised by the economists.

It might be thought that a flying flock of fattening sheep would be better suited for a small farm such as this, and in some respects this is true; but nothing would compensate us for the pleasure of seeing the lambs playing in the spring sunlight, running races up and down the banks of our clear-flowing stream, leaping over it, and getting such fun out of life as only lambs can. How we look forward to each lambing season, with its promise of spring; how we enjoy winning the confidence of the ewes and being able to identify the lambs which belong to each. How we curse pet lambs—orphans—which have to be reared by hand, knowing they will never pay for the time taken; but we haven't the heart to cut their throats, and feed them on warmed Jersey milk every two hours. This Jersey milk seems ideal for the purpose, the lambs being equal in every respect to those reared by their own mothers, in striking contrast to the miserable, pot-bellied little runts which one sees on so many farms where bottle-reared lambs are kept.

Our elaborate lambing pens, built of bales and hurdles; our concrete sheep dip, with its roller to put in the sheep gently and the gate to regulate their passage through the dip; the sorting yards and draining pens, the footrot trough, power shearing plant, and other refinements, all for forty ewes, are a source of amusement to some of our neighbours, who run big flocks on their great arable farms; but we do like to have a complete unit in anything we undertake. Anyway, we are compensated for our trouble when the lambs are sold and the auctioneer says, as he has done a number of times, 'Now, gentlemen, we come to the Oathill consignment, and the best you will see here to-day'.

Sheep do mean a great deal to us, and especially to me, for it will be remembered that I was trained as a shepherd; and I always say that when, in common with the rest of the poor farmers, I go bankrupt, shepherding is the occupation I will follow. For the sight, sound, and smell of these animals never fails to stir me, and the finest sight I know is a great flock grazing on a green mountainside. On the other side of our valley rises a steep bank on which a neighbour grazes his flock, and as I go about my daily work my eye constantly roams over them, and if there is very little in this world he has to thank me for, I do not think he has ever lost a cast ewe between dawn and dusk since we have been here; for in my mind's eye I still see the great flock which was my first real responsibility in the days of long ago.

But until the day when I do again have to start working for a living (for farmers, so I have been told on the best authority, live by exploiting the workers) I can at least have the pleasure of training good sheep dogs, and with the possible exception of riding a good horse, there is nothing I enjoy more. Just as in riding a horse one co-ordinates the equine muscles with the human brain, so one employs the inherited and natural aptitudes of the sheep dog to execute one's commands.

There are several ways of training a sheep dog, such as tying it to an old trained dog, or running it round a field on a long light string, until it will obey the words of command, but in my opinion these are clumsy methods and you do not get a polished performance; the dog is either looking for his companion, or abuses his liberty when no longer on the string. The best means I know is to have a well-trained flock of sheep, for contrary to the common opinion the sheep is by far the most intelligent of all the farm animals, being superior to pigs, cattle, and horses, in that order, and only has to yield to the dog, who is backed by the human brain.

As you go into the field with a well-trained dog at heel, the sheep gather together in the middle of the field and stand facing you, for they know that that quick dark shadow will appear on the other side if they attempt to move off. When a farmer says that his sheep go out the other side of the field when he enters, you may be certain they have never been used to a trained dog. In this way you may inspect them, and handle them if necessary, for the good dog can hold sheep in the centre of a field, without taking the flock into a corner, for it must be remembered that a mountain shepherd has no hedges or walls in which to pen up the sheep, and only the lowland farmer needs to resort to the other method. It is on this gathering together in the middle of a field, with the help of a well-trained old dog, that I rely in training a puppy.

As I approach the sheep, they gather, and the puppy with his inherited instincts is all a-quiver with interest. He moves forward a few paces, I say 'Go'. He stops, and I say 'Stop'. A sheep puts down her head at him, for he is not the familiar dog; he draws back, I say 'Back'. He moves round to his right, or to his left, I give the appropriate order simultaneously. As soon as he tires, I reward him and take him home. In a few days he gets a little bolder and will perhaps run at the sheep, but being well trained themselves they only gather closer, and the puppy, wondering what to do next, looks to me, and the familiar word will send him scampering round to the right or left, to draw back or creep forward. And so it goes on, a few minutes a day for several weeks; the play of his puppyhood grows into his serious work for life, for a dog, like a man, enjoys doing that which he has been thoroughly taught to do, and it is only neglect, ignorance, or carelessness on the part of its owner which makes a dog a nuisance to itself and everyone else.

With ten words mastered a good dog can manœuvre the flock into any position you may desire for the rest of his life. Those I use are, 'Go, Come, Steady, Stop. Go round (he turns to his right). Come round (he turns to his left). Down, Heel, Yes, and No'. The last two are used in sorting out a flock, when he has to let go one sheep and keep back another, or to encourage or reprove him at other times. In working more than one dog I use each order in conjunction with the dog's name, such as, 'Stop, Wisp', 'Go, Fly', and it will be noticed that single-syllable names are chosen for the dogs, so that they never have to master more than two sounds to carry out an order. Some people use French, German, or Gaelic for a second dog, but it is quite unnecessary, for a dog can easily associate his name with an order. Also, in selling a dog, a new owner can seldom be bothered to learn ten words in a foreign language, and if he is only going to work one dog, then the name can be dropped, as it no longer serves a useful purpose. A whistle can be used, especially if one wants to use a dog at very great distances, but it is not every one who can whistle ten clear and distinct notes, and it takes twice as long to train the dog, and some never learn; possibly, like human beings, they do not all have the same ear for music, although many of them can hear notes which are inaudible to us.

In normal times a good, well-bred puppy can be bought for a couple of guineas, which is cheap if he is required for one's own use, but few farmers will give more than a fiver for a trained sheep dog, and then, often as not, they will spoil a year's careful training by a minute's bad temper. I once sold one, giving the man a demonstration on his own farm, and very carefully explaining the orders to be given, and the response he would get from them. The next day he took the dog out, and it worked well and he sent me a cheque. By the following day he must have forgotten my instructions, for he thought the dog would work to the owner's right and left, and not to his own; for when the dog faithfully carried out his commands as given, he took them in the wrong direction. The exasperated farmer used a lot of bad language, which confused the issue, threw a stone, which brought the dog to heel, where he had always felt safe during his training, to be rewarded by a kick in the ribs, which convinced the anxious and willing creature that he must no longer obey his new master, whatever words were used, and he refused to work again. Naturally the man wanted his money back, which he got, and it took many weeks to restore the dog's confidence in human nature. There are, of course, many men who can be trusted with a good dog, but these usually prefer to train their own, and only if they must have one immediately will they buy. So generally speaking, like so many other things in farming which are so worth while, the training of sheep dogs is not an easy way of making money. For many years now I have made it my rule that I will only train one for a friend, and only then if he is not given to swearing, a sure sign of that instability of temper which is fatal in dealing with man's oldest and most faithful friend. Also one must make sure that he has not a wife who will pet and overfeed it, which has been the ruination of many a fine dog, and when he slinks off back to the kitchen door as soon as he has finished his work, you may know another has fallen as did Adam in the Garden of Eden. For a good dog would rather work than eat any day.

But whatever the perversity of human nature, women in particular, and the world in general, the shepherd has one consolation; he can always go and look at his sheep. Unchanging in a changing world, still counted in the numbers of a lost language hidden in the mists of antiquity, still breeding true to the signs of the stars, still the symbol of purity and love. And if the lock of wool is no longer placed in his coffin, as by the medieval monks, as his passport into heaven, he finds his paradise here on earth if all is well with the flock.

# CHAPTER SIX

## THE PIGS

In our early days at Oathill there were only three small old-fashioned pigsties, not suitable for breeding sows; so we contented ourselves with buying good litters of eight-weeks-old pigs and fattening them out. It was our rule never to give more than £1 a head for good weaners at this age, which was the price we could have bred them for, with a small margin of profit, had we had the accommodation available. If little pigs were making more than our limit we left them alone, for we knew our cost so well that a shilling extra on the price we paid would reduce our profit in proportion. In a boom period little pigs would make as much as fifty shillings each, to lose a lot of money by the time they were fattened out and sold in the inevitable slump. For before the establishment of the Pig Marketing Scheme in the early nineteen-thirties, there was a regular trade cycle in which pigs were copper or gold, muck or money. If fat pigs were dear, farmers rushed in and bought in competition with each other, forcing up the price of store pigs. Others started breeding, for the pig is a most prolific animal, and here was easy money. In a few months down tumbled the prices, due to the glutting of the market with fat pigs. The fattener lost money, the breeder had pigs left on his hands, which he was compelled to fatten, or cut his losses by selling in a depressed market where there was no demand. We have seen good pigs sold for sixpence each, which a year earlier would have realized thirty shillings. Fat pigs vary from twenty-six shillings a score, down to ten shillings. The vicious circle was hastened by another factor; the demand for feeding-stuffs, due to an increasing pig population, caused a rise in costs which cut into rapidly reducing margins of profit, and added to the troubles of the poor pigkeeper.

And yet there was a profit to be made in pigs, for we averaged £1 a head over a period of three slumps and booms, and in the whole of our experience we have only lost money on one bunch of pigs; yet there were many old and experienced farmers who told us, and we had no reason to doubt them, that they had lost as much as £1,000 on pigs in a year. Too many, hearing that bacon pigs were selling well, would decide to fatten some, if for no other reason than that cereal prices were tumbling and they had barley on their hands. They would go to market and study the trade. Noticing that useful store pigs were averaging perhaps £2 apiece, they would consider an excellent bargain had been made if they could be obtained at less than that figure, for there is no-one to equal an old farmer in making a shrewd deal. And this would of course be good business if the pigs were to be sold again immediately at the market price, but the subject has to be studied from a different angle if the stock are to be fattened out.

First one should think, what will these pigs be worth when they are fat? How much food will it take, and how much will that cost, to fatten them? What profit do I require? And then, therefore, what can I afford to pay? How simple this sounds! Yet I have never met a farmer who works on these lines, and there are very, very few, who can tell you average consumption of food on their farm required to produce a score of bacon, the unit of twenty pounds dead weight on which the fat pig is calculated. Far too many also make the fatal mistake of thinking that if others can fatten pigs profitably, they also can do so, for not only is careful buying necessary, but the consumption of a pound of meal per head per day more than the pig requires, will eat up the entire profit, so management enters largely into the matter, however shrewd a buyer one may be.

Our method never varied. Buying only when good stores could be obtained for twenty shillings or less, we fed regular quantities according to age of carefully balanced rations, so that we always knew our costs. We sold as soon as we could see £1 a head over and above the cost of feeding-stuffs and weaners. It might be at six score, when the trade was good, early in a boom, or run on to ten score in a failing market. To sell at five score the initial outlay on the weaner is four shillings per score, while at ten score it is only half that. When we could sell at the lighter weights, we could fatten three lots in the year, but with heavy weights only two. So that pigs at all times gave us the quick turnover of capital, which we contend is the first essential in profitable farming. Our three little sties, with a capacity of about twenty pigs, filled two to three times a year, and only taking up a little time, could bring us in £40 to £60 a year, while large-scale pig farming was one of the quickest ways of losing money, as many found to their cost and disappointment.

It was also our rule to fill the sties and fatten out each bunch before we bought again, so that if we should inadvertently buy any carrying swine fever, we should only experience a comparatively small loss, and not infect nearly fat pigs. This would have been a disaster for us, with four or five pounds invested in each, and probably carrying a little credit in their last month, on the same system described in our poultry section. Fortunately for Us, we never experienced trouble in this respect, though it is said that those who buy in the open market sooner or later buy disease, and we were very much afraid of doing so. We think our system was well justified, for we have known others incur great losses by buying a litter every week or month, and then swine fever or erysipelas has swept through the stock. And although there may be a margin of profit for careful buying and good management, there is none to cover losses from disease.

As was not the case with the other stock on the farm, our £50 invested in this branch was sufficient to stock to the capacity of the available buildings in the first year. Twenty pounds would buy twenty pigs, thirty would be laid out on feeding-stuffs, and with £20 worth of credit skilfully worked for the last two months of feeding, our pig capital increased to nearly £90 by the end of the first year. As it could not be reinvested in pigs, it was lent to the general farming account, which was always in need of money, until such time as it could be used again for increasing the herd, for with pigs, as with other stock, we determined to have pedigree stock only.

At last the time came when we could see our way to build a large Danish-type pig-house, and found the Enstone herd of Pedigree Large Whites. Over the years we were waiting, we visited as many pig farms as possible and studied their buildings. While we appreciated the saving in labour with the big corrugated-iron pig-houses, we considered them much too hot in summer and too cold in winter. A fattening pig will take as much as a hundredweight of extra food in severe weather just to keep itself warm, without any increase of weight. If the weather is too hot, then the pig receives another setback, and appears especially susceptible to disease. While with perfect conditions, in a suitable building it is possible to fatten pigs out to eight score, eight pounds, at seven months old, on six hundredweight of meal, with factory-like precision, or 210lb. live weight in 210 days. This was very important, when under the Pig Marketing Scheme one had to contract to supply a certain number of pigs per month, at a given weight and grade, if the best price was to be obtained.

We built our pig-house, described by some who have seen it as a 'pig-palace', of wood and asbestos, with a wallboard lining, leaving a three-inch cavity between, ensuring as even a temperature as possible at all seasons. A hollow floor also ensured warmth, a most important feature in pig-house construction. Water was piped on to every point required, both for feeding and swilling down. Later we arranged our steam boiler so that steam could be driven through all the water pipes to prevent freezing in zero weather. A large liquid manure tank was also incorporated, for we looked to our pigs to return to the land in full measure the straw and green crops they received from it, and we hate to see potential profits running down a drain. Pigs should be muck and money, not muck or money, as the old farmers used to say.

Buying six pedigree Large White gilts (for so the young female pigs are called) and a boar, all tuberculin-tested, we had a flying start. Having built our reputation with poultry and cattle, people seemed to take it for granted that our pigs would be up to the usual standard, and we certainly did our best to see that they were. We had many advantages which the specialist pig farms had not: milk from the Jerseys to give the weaned pigs a good start when taken from the sows; fresh green, kale, rape, vetches, and mangolds from the arable; ample homegrown straw to ensure their comfort; and a run on fresh grass for the breeding sows.

By the end of the first year we had increased our stock to twelve sows, for this was the number, with their followers, we had designed the house for. It was so placed and laid out that we could extend if necessary in the future, but all we required was one complete unit, to fit in with the general balanced system of the farm.

In the second year we were able to sell fifty gilts and twenty-three young boars for breeding, from the two hundred pigs we bred from the twelve sows. Surplus pigs were fattened out and realized satisfactory prices, although at that time many farmers were grumbling bitterly at the price which had been fixed.

Under the scheme pigs were graded into four classes, 'A', 'B' 'C', and 'D'. The first and second grade left a fair profit, the others did not, for not only was the price low, but the grade was a sure indication that the pigs were of the wrong conformation and probably overfed, or run on to too heavy weights.

It was considered that 70 per cent grade 'A' was good, and several pig farmers wrote to the farming papers when they achieved it, but we still have the grading sheets which show our own pigs came out at 97 per cent grade 'A'. I say own, for I am convinced that on two occasions the return we received was not for our own pigs, in spite of the assurances from the factory that they had a completely foolproof system and never made mistakes. In one case they showed two rigs, which are pigs which have been only half castrated, while it is our unfailing rule to locate both testicles before removing either; this proved to our satisfaction that they were not our pigs. On another occasion they paid us for a score per pig heavier than those we sent in, as weighed over our weighbridge, on which every pig is checked as it goes to the lorry. In common honesty, as we did not want some other poor farmer to lose his due, we drew attention to the discrepancy, but once more they swore by all the red tape in officialdom that it could not happen. So there we had to let the matter rest; what we lost on the rigs we gained on the weights, but that does not happen to every individual farmer, for under certain circumstances he might lose both times.

There is no doubting that right conformation, obtained by careful breeding, is necessary to obtain a high percentage of grade 'A' pigs but management is all-important. We feed five times a day, between 5 a.m. and 9 p.m. for the first month after weaning; and then every eight hours for the rest of the fattening period. With small feeds and limited feeding, we never give more than six pounds of meal per day; the pig's stomach is never distended, with the result that a high-grade carcase is obtained.

To feed at so many regular intervals, someone might inquire if we lived with the pigs, or had any other work to do, but in fact it fits in well with the ordinary routine. Using the swinging flap over each trough, so that the next feed can be put out as soon as the first is cleared up, it is only necessary to walk down the house and release the catches for the pigs to help themselves. If two feeds are put out in this way first thing in the morning the pigs have their second feed mid-morning, when the house is cleaned out; and another feed put out then involves only the releasing of the flaps at midday. Then two feeds put out at the ordinary evening feeding time enables the master, in his final look round at night, to feed the whole lot again as he walks down the house.

I remember reading of some experiments at the Harper Adams Experimental Station, in which they did not get any better results by feeding three times a day. But they studied the convenience of the pig-man, inasmuch as the pigs were fed between 7 and 5 o'clock. Had they fed at 5 a.m., 1 p.m., and 9 p.m., I believe they would have found different results, as this has been our experience. The time involved in multiple feeds is no greater; we found two feeds per day would mean half an hour each time, while taking round a small quantity three times only took up twenty minutes, morning, noon, and night.

Under war conditions pigkeeping brought many new problems, though they have been overcome, and it is still a profitable branch of the business, while the pig manure is more valuable than ever for our light land under the strain of war production.

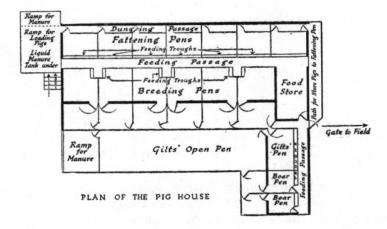

PLAN OF THE PIG HOUSE

Quite early on we adopted the single litter system, under which a gilt farrows once (the finest of her brood being kept on to breed the following year), and is fattened out. This enables the herd to be kept very cheaply through the summer, while no little pigs have to be reared in winter. Latterly we have modified the system, by keeping the best gilt, as judged by all the litters produced, to go on and breed again, as stock from a sow is generally supposed to be better for breeding than that from a gilt. In this way a very high standard of selection can be maintained, ready for the cessation of hostilities when the pig breeder will come into his own again. After the last war we saw sows of no particular merit realizing over £100 apiece, and while such inflated prices do more harm than good in the long run, if any money is being thrown about we should like to have our share, to tide us over the slump that follows.

From the aesthetic point of view pigs do not appeal to everyone, though it is very pleasing to the farmer's eye and mind to see a house full of contented, thriving pigs. That the pig is the cleanest animal on the farm, given suitable conditions, is now generally well known. Conditions of working can be clean and pleasant, far more so than in many cowsheds in winter where the animals lie in, and pigs probably lead a far pleasanter life than the majority of farm animals, so there is a lot to be said for pigkeeping. Some people get really enthusiastic. One evacuee girl we had early in the war thought them horrid the first time she saw them; but in a few days she took to feeding them regularly, until one day she was invited out to tea, and refused on the grounds that it would mean missing feeding the pigs. She was accused of loving the pigs. 'No,' she said, after due reflection, 'I don't think I love them, but they have a pathetic appeal, like slum children, only much nicer.'

## CHAPTER SEVEN

---

## THE LAND

However profitable and interesting the stockbreeding may be the land remains our abiding interest and first duty. The stock, like us, will pass on, the land will remain.

The cleaning and reclaiming of this poor, neglected soil, mentioned in our plan, has been well justified, though of necessity a long-term process; and if at times there has seemed little return for the thought and labour put into it, time has proved it well worth while.

The cleaning of one-fifth of the arable each year by repeated ploughing involved the walking of fifteen miles per acre, with an eight-inch furrow (for this was as much as could be turned to the full depth in the matted, stony soil), each of the six times it was ploughed. On eleven acres this means something like a thousand miles between spring and autumn. Over every inch I drove our team, while my brother devoted his attention to the hundred and one jobs which needed doing between the times we had to devote to haytime and harvest, and the pulling of weeds which infested our corn crops. The stock, of course, was tended before and after the hours the horses could be kept at work.

After the first harvest we were in a position to increase our horses to four. Then I could take one team to work in the morning from seven o'clock to noon, the other pair from one o'clock to six. Thus I was able to push them along harder, while a spare horse could be used in the afternoon for odd carting jobs. Also by working the four together we could use a cultivator and tear out the weeds to dry on the top before being ploughed in again. In these days of powerful and heavy tractors many of the younger farmers are apt to forget the work and hours which had to be put into the land when the horse was the unit of power.

Our belief in the destroying of couch grass by ploughing in has been more than justified, for it has completely disappeared from our arable, enriching the land in the process. For it must be remembered that so rich is the root of this weed in plant food that it has been dried and ground for human consumption in times of scarcity. In our experience the burning of a heavy crop of 'squitch', as the farmers call it, robs the soil more than the taking of a full crop of wheat or other cereal, while leaving behind the tiny pieces of root to start all over again.

In the autumn of 1925 we were able to buy a second-hand tractor for £30, which put us right on top of our work. While we still used the horses as much as possible the tractor was able to make up arrears of work in late seasons or wet times, and our early years were, generally speaking, wet years, in which on the majority of the farms the fallows got out of hand, and one would see weeds going to seed in the root crops—a disastrous state of affairs, for this is the only opportunity the farmer has for cleaning the land over a period of years.

In the second course of our rotation, that is to say after the first five years, we were able to grow root crops, chiefly rape and kale, but some mangolds were grown for late feeding. The latter are invaluable if a heavy crop can be grown and carefully stored. In the drought of 1943, we were feeding mangolds from the previous year to sheep and pigs in mid-August.

Soiling mixtures of oats and vetches have always appealed to us. In the early years we bought vetches as cheaply as four shillings a bushel. We cut them green for cattle and horses, penned off the second crop with sheep, or made them into hay, or if the growth was too luxurious, and there was more than could be used, then we ploughed them in for green manure. It is rather interesting that even a thick young heavy crop of vetches carefully ploughed in has never shown any increase in the following crop. This is one of the mysteries of farming, for we have always good results from clover or mustard, and are firm believers in green manure.

It is sometimes said that a man only makes one crop of good vetch hay in his life—this crop requires so much drying and is easily spoiled—but actually on four occasions it has proved our salvation; for although this farm has been very heavily stocked, we have never had to buy hay since 1926. We do recommend vetches with every confidence for they give a heavy crop of high-quality hay, smother weeds while growing, and are harvested early enough to give a thorough cleaning to the stubble, or if rye grass is drilled in, or even mixed with the seed, will give a good second crop for grazing. In one very wet summer, 1924, we took two hay crops, one in June and the other in late August, beating the weather on both occasions by a few hours; for one wet day depreciates this crop by fifty per cent. If one is prepared to use twenty pounds of agricultural salt to the ton, vetch and oat hay can be carried quite as quickly as a clover mixture, for this prevents any tendency for heating to take place in the stack, while the finished product retains its green tinge, and cattle prefer it to the ordinary hay. If weather is very unsettled, vetches can still be harvested by cutting small areas, cooking as soon as possible (tripods would be ideal, though we have never tried them), and spreading a cut-open bran sack, weighted at each corner with a stone, over each cock, for the open nature of the crop prevents the shedding of rain to the same extent as with ordinary hay. The sacks of course can be used over and over again, and make excellent chaff sheets for use in winter. A really heavy crop of vetch and oat hay will come out at three to four tons to the acre, very useful when a late spring or a dry spell reduces the yield of ordinary hay.

*OUTSIDE YARDS FOR BREEDING-PENS*
*FATTENING PIGS: INTERIOR OF THE PIG HOUSE*

*THE SOIL (NOTE THE STONES)*

On the subject of hay-making we hold rather strong views. Having learned most of my farming in the wetter and colder north, I have always viewed with disfavour the south-country methods of haymaking, in which the crop is either sun-bleached or rain-washed according to the season, to say nothing of the most nutritious parts being battered off by swath turners, hay tedders and loaders. I was taught that even the horses must not be walked across the cut swathes, while here many farmers walk them up the windrow itself when using a hay loader; though in recent years the use of tractors, which of course straddle the row, has brought about an improvement in this respect.

Unless the weather is set fine and dull, when we sweep the hay straight to the rick, we believe in putting it up into large weatherproof cocks, containing about half a ton, so that we get green hay undamaged by rain or sun. It may take a little more labour, but this is more than paid for by the extra quality, for good early green hay is one of the cheapest foods on the farm, and stores the best of all. And the brittle, sun-dried product, from which all the leaf and seed has been battered, is in our opinion, of little more value than good oat straw.

It is probably no exaggeration to say that many farmers waste a quarter of their hay in the field, as shown by the analysis between samples, and another quarter down the cow's throats for want of careful rationing. Too much hay reduces the production of milk by overloading the digestive system. We believe better results can be obtained by feeding only five pounds of really first-class hay against fifteen or coarse and half spoiled. For rationing, we rely on baling, for even on this little farm, one pound of hay per day per beast more than is required involves a loss of £40, at present prices, during the winter. What must it be over the country as a whole?

Generally speaking, by cocking wilted hay at the earliest possible moment we only require half the fine weather necessary to the other method, and it is safe through long spells of rainy weather when so gathered. We have had an inch of rain in a day on cocked hay, but a good drying wind following has enabled us to go on sweeping it into the rick twelve hours later. Any extra work is more than justified, when as in 1927, the wettest year on record, our postman told us we had the only rick of good hay on his whole round, nine-tenths of all the hay in the district being burnt in the field to get it out of the way after it had been completely ruined by the weather. In twenty hay-times I think I can truthfully say we have never spoiled more than a couple of wagon-loads of hay, while by careful rationing we have fed it to the best advantage and the last handful. How often we have known farmers to buy hay at famine prices in late April because the grass has not come, when saving a stone a day over the whole winter, which their cattle would hardly miss, would have given them a reserve for a late spring. As far as possible we like to carry a reserve of hay from year to year, for a good stack of hay will justify the rent of the land on which it stands when the lean year comes, apart from being better feeding quality by storing. If racehorse trainers insist on old hay and oats, surely the same must also be better for more useful stock?

In growing corn, we made it our rule right from the start, that only the very best would be good enough. When ordinary seed from the local merchants could be obtained for £3 a quarter, we went in for the very best at twice the money from one of the leading seedsmen. This to many people would seem extravagant, for farmers starting with the minimum of capital. But to us it seemed fully justifiable; in the first place it was possible to obtain twelve months' credit, by the kindness of the local agent, who put our case to his firm; this was worth a lot to us. Secondly, we knew then, as we do now, that the best is the cheapest in the long run. One should consider price per acre, not price per bushel, for only half to two-thirds of really first-class seed will give as heavy a crop as any seed which has only cheapness to recommend it. Also when grown an enhanced price can be obtained, for even once-grown seed can command its price against that of unknown origin. It has always seemed strange to us to see large-scale wealthy farmers planting perhaps only an acre or two of some good variety, to provide their own once-grown seed, when they could make several shillings a quarter extra on their whole output by growing only the best. It always paid us to sell all our corn for seed, and replace it by locally grown wheat and oats for stock feeding. In the steady decline of prices during the depression we always threshed as soon as possible after harvest, selling it for seed, and then perhaps buying at ten shillings a quarter cheaper a few weeks later on a glutted market. Possibly the most striking example of this was in 1927, when our corn cheque came to £333, and we bought an equal weight of corn back later for £197.

After a few years we were lucky in obtaining a contract to grow direct for a seedsman, he to provide the seed, while we received an agreed percentage over and above the market price. Special precautions were necessary, and facilities had to be provided to enable the crops to be 'rogued', i.e. any ears which were not true to type to be picked out. This is only one of the things the farmer pays for when he buys good seed. With perhaps three-quarters of a million heads to the acre every one is sighted as the experts move carefully through the crop. Very little damage is done, while the heads removed can be used for poultry feeding. As a rule only perhaps a bushel will be found in twenty acres, but they have got to come out. It is really surprising, too, what a lot can be learned from these men who spend their whole lives in this class of work, and we are always grateful for any scraps of knowledge which can be picked up, and which can be used to obtain greater efficiency on our farm, or passed on to those who can benefit. To look at a crop of corn one needs insight, as well as eyesight, and there are half a dozen things one can see in a single stalk of wheat which might easily be overlooked in a whole lifetime on the land.

The careful compiling of records over our twenty years' farming has brought out many interesting features; careful recording is of far greater importance than the majority of farmers realize, and well worth the little trouble involved; for, if nothing else, records compel observation, which is essential in farming. Recording leads one to experimenting, and although it is sometimes said that a farmer cannot afford to experiment, we have never regretted doing so. How easy it is to try different dressings of manures, different seed mixtures, different rates of sowing, and other departures from the accepted practice. For example the standard rate of drilling oats in England is four bushels to the acre. We tried six, which is the custom in many districts of Scotland, and reaped an extra twelve bushels to the acre on heavily manured land, but only stunted the straw when we grew with artificials alone (when the dung could not be spared). Probably natural manure retained sufficient moisture in the soil necessary for the extra growth, which was otherwise lacking in this comparatively dry climate.

The average yield of grain has steadily increased over the years from seventy quarters to over two hundred in the years between 1939 and 1942, while the grazing stock we can carry has increased in like proportion, due very largely to that most neglected and discredited substance—muck. How often we have been told, 'Muck carting does not pay'. Yet we spend more time on this 'unprofitable' work, some six hundred hours per year, than any other operation on the land, which would indicate that something else must be very profitable indeed to justify the figures given in our chapter on accounts, if muck carting is uneconomic. Personally I have no doubt that we have gained more by the careful conservation of manure and its proper application to the land than we have ever received in subsidies and other Government assistance during the last twenty years. This, I think, justifies the belief constantly expressed in this book that the solution of any difficulties the farmer has lies in his own hands, so they need not be used for holding out his hat to the taxpayer. Much has been written about winning the sympathy of the townsman; nothing would do that more thoroughly than efficient farming, and it would then not be needed, for farming could stand on its own feet like any other business and supply the food required at reasonable prices.

There is very little that is new or original to say about the value of carefully stored manure. Most of the early writers on agriculture appreciated its value and the importance of protecting it from rain and sun. Manure trodden in open yards is in my experience practically valueless; most of its goodness goes down the drain. My attention was drawn to this in 1922 when, in looking at a field of oats with an old carter, he mentioned that it had been half mucked the previous autumn and you could see to the last cartload where the manure had been spread. This was the first time he had seen any response on that part of the rotation for many years. I pondered on this, and suddenly realized that 1921 had been a very dry winter and spring, so that the manure had not been rain-washed and was therefore equivalent to that made in a covered yard. I knew that Scottish farmers, who in the east of that country at least invariably have covered yards, always looked for good results from an application of manure, and my inquiries showed that this was so.

The best manure, I was given to understand, was made by littering the animals daily, with a thin layer of straw, not by drawing in a cartload once or twice a week, which is the more common English practice.

As soon as we could we built two covered yards for the young stock. For the cows (we usually only have four or five in milk) we still use an open yard, but only the part where they lie under cover is littered. Any droppings in the open are shovelled up daily and thrown under the shed before littering. We only use the cowshed for milking, so there is no loss of urine down the drain. The cows lie clean and comfortable in their open-fronted shed, while their manure is preserved to benefit the farm; a strong case can be made out against the milking cow as the robber of the land when large quantities of milk are sold off a farm, and valuable by-products are swilled down the drain. We use sufficient straw to absorb all the liquid, about two hundredweight per day.

At the pig-house we have a large concrete liquid-manure tank built in. Large quantities of peat moss are used for the poultry, and this being seven times as absorbent as straw, when carted out is soaked to its maximum holding capacity with liquid manure, something like two hundred gallons to the ton; this never fails to show its value as a top dressing. It is far more valuable and more lasting than a heavy dressing of sulphate of ammonia, the great standby of the British farmer, which compels the crops to steal from the land the depleted stocks of potash and phosphates, to say nothing of the rarer but equally important substances not yet fully explained by science, which are of greater value than is generally recognized.

There can be no doubt that the yield of grain and length of grazing season have steadily improved year by year on this farm as the humus content of the soil has increased, while we have never seen the slightest trace of wireworm damage, which is quite common on neighbouring farms. We believe that the wireworm are so busy fulfilling their proper function of reducing humus to plant food that they do not need to attack our crops. You can find them by searching the soil, but not by pulling up a plant by its roots.

It is rather remarkable that two out of three farmers walking round the farm, under the War Agricultural Committee's demonstration scheme, will inquire if a particular crop is grown after ploughed-out grass, owing to the strength and quality of the straw. They have learned to associate fertility, since the war, with the accumulated reserves of plant food in old grassland, in contrast to exhausted arable. They are usually surprised to learn that we seldom have more than a one-year ley, and that the good crop is the result of farmyard manure.

On nine out of ten farms the best field is the one nearest the buildings, for the simple reason that it has had more than its fair share of muck. We have now made our most distant field equally good, or a little better, by always starting mucking that side of the farm and working back towards the buildings, as far as the rotation will allow.

We make some six hundred cartloads per year, enabling us to muck about half the arable. The ley is penned over with poultry-fold units, either before or after the hay crop. This alone leaves behind the equivalent of four hundredweight of sulphate of ammonia, two hundredweight of superphosphate, and one and a half hundredweight of kainit, in, we believe, a more valuable form, for the results show for two years; ammonia first year, phosphates the second.

We always spread manure direct from cart or trailer, never putting it out in heaps, and as far as possible plough in each evening all that has been spread during the day.

From August to April we can usually cart direct from the yards to the land; during the summer it is sometimes necessary to make a manure heap. As far as we know no special precautions are necessary to prevent losses in the heap in a normal year. At one time we used to dig a narrow trench all round the heap, but finding that drainage was negligible we discontinued the practice. The heap must be straight-sided and square to prevent undue fermentation; normal summer rain helps by keeping the heap damp. Season and circumstances permitting, we sometimes plant marrow seeds all over the heap. The strong vines and large leaves keep the heap moist, while we reap a useful crop of heavyweight marrows from July onwards.

We are sometimes asked if we make any compost. We have nothing to compost. There is no waste vegetable matter; anything edible is made into silage, weeds and potato tops are ploughed. Mud scraped up on the road is put in the covered yards with straw, or carted direct to the land. Possibly manure made in a covered yard is the perfect compost heap, when straw is spread every day, and trodden in with cattle, in sufficient quantities to prevent loss by drainage.

We recently had the pleasure of entertaining an Indian student who was taking a keen interest in modern English agriculture and we asked him how our methods compared with the best Oriental practices. He expressed the opinion that the underlying principles were the same—returning everything possible to the land, with the smallest loss and as quickly as possible. The law of the forest was being faithfully obeyed, inasmuch as even the leaves from the hedges were ploughed in; though the student deprecated the loss of land, about one per cent, taken up by hedgerows. He modified this view somewhat when it was pointed out that by providing shelter for bird life it helped to maintain the balance of nature. Also the hedges provided shelter for stock, wild fruits for jam- and wine-making, poles and rails for fencing and other light work, and served as snow breaks and checked land erosion by wind. Those who would fit the English countryside to the tractor have never pondered the reason why our forefathers planted their hedges, and studied on the spot, asking themselves, 'Now why was this particular hedge planted here?' It is quite a fascinating hobby, for you can nearly always find the reason, and you realize that the old farmers were not nearly so foolish as some people imagine.

Our Oriental visitor was impressed by the fact that we never use poison sprays. The best preventative of charlock we know is a thick, thriving crop, while the necessity for spraying potatoes has never arisen on our farm. He also expressed the opinion that, while our arable farming compared favourably with the methods proved over thousands of years in his own country, the stock here was far superior.

With the simplicity of the foreigner, he inquired if we were yokels. He was assured, indubitably. Why? He had gathered from his tutor in Oxford that agricultural work in this country was performed by a lower order so designated, but in talking to the young men employed on the farm (actually pupils and the sons of professional men) he was impressed by their intelligence and physical perfection, comparing favourably with university students, who, he understood, were the flower of English manhood.

I relate this story without any class prejudice, which to me indicates that those who would re-educate the great British public to the importance of agriculture have quite a big task before them, if university teachers still think their food is grown for them by yokels, known as 'yawnies' on the Cotswolds, and 'chawbacon' (from their staple diet) elsewhere.

While being great believers in natural manure and humus we do not condemn artificial manures; they are useful for replacing the loss due to the sale of eggs, flesh and bone in the stock sold off, and the corn we are compelled to sell in wartime. Losses due to the sale of potatoes, another wartime necessity, are more than made up for by the consumption on the holding of pig potatoes bought in large quantities to maintain the pigs and poultry. Our aim is always to build up and maintain a fertile soil, as the basis of health in crops, stock, and human beings. For they are very closely related, far more so than is generally realized. I was once told that the Romney Marsh, one of the most fertile sheep-grazings in the world, was made by keeping ten sheep to the acre until each acre would keep ten sheep. The same principle has proved true in our farming.

This is illustrated by the extreme measures some owners take to ensure that not a single hen shall trespass on their land for fear that it might carry infection with tuberculosis to their attested herd of cattle. On this farm we run a hundred birds to the acre where the cattle graze, and all their food is grown with poultry manure. As a result, we had two reactors between 1924 and 1942, both of which afterwards tested clear when the man who bought them as reactors decided to test. Few breeders could show better results over a longer period for a similar number, on land from which every hen has been excluded. It is, of course, possible that it is the little daily dose which builds up immunity. But this we must leave for the scientists to decide; we are content with the practical results, for even when the new Weybridge test was brought out, which was said to be so searching that every attested herd in a whole country had its reactors, ours still tested clear.

Early in this century vitamins were discovered—the vital food factors which ordinary analysis does not disclose, though the value of certain foods for a specific purpose was known; perhaps in a hundred years' time similar substances will be isolated which are equally essential for the maintenance of health in the soil. Till then we are content with muck, just as the old-fashioned doctor who had never heard of vitamin D recommended cod-liver oil for rickets.

I have a book published in 1572 in which the author suspects that there were some essential salts in manure which benefited the crops, but which could be lost by careless storing. Perhaps in three hundred years' time some writer in the future, though possibly he will simply record his thoughts by some electrical process instead of forming half a million letters with a pen on paper, will come across an odd copy of this book and think that I too was on the right lines, however quaintly I express myself in old English, while he, of course, uses the common international language introduced in the twenty-first century!

Even with a fertile soil thorough cultivations are still necessary to make the plant food available to the crops. I would not indulge in such a platitude as this, had I not read in the agricultural press a few months ago of members of the Farmers' Club solemnly going up to London on a fine afternoon to be told the same thing by an eminent scientist. Their local gravedigger could have told them, and saved a journey that was not really necessary, when every hand could be more usefully employed on the land.

What a mistake is made, rushing over the land with the minimum of cultivations, distributing artificials haphazard with badly set drills—I have yet to see one that could equal accurate hand broadcasting—and calling it mechanized farming! The real value of the tractor is to enable more thorough and careful cultivations to be made in a limited time, not simply to speed up the old methods. The real test is that the tractor must increase the efficiency of tillage operations so that moisture is conserved in dry times and assisted to percolate in wet weather. Mechanical power should never be regarded as labour-saving unless it is associated with increased production, for it must be remembered that the Roman Empire declined and fell when the peasant farmers were replaced by slave labour (which is a form of power farming) and the resulting fall in production made the country dependent on imported food; until finally, when unable to maintain its trade routes, the whole Roman system collapsed. The same thing could happen to-day; if Britain were unable to import tractor fuel, the people would starve in a war simply for lack of sufficient capable peasant farmers and a fertile soil. Large-scale farming in this country is the writing on the wall. When the guts have been torn out of land stimulated by artificial manures, robbed by selling off all the produce, paying income tax on sales of straw which should have been consumed on the holding, then the farmer will find himself in the same position as his father after the last war; declining yields and prices will complete his ruin. Who can take warning from a fool, who values his muckheap higher than his bank balance?

# CHAPTER EIGHT

## LABOUR

Probably the biggest problem farmers have to face is labour. With enough good and efficient men a really capable farmer could manage a thousand acres as it should be farmed.

The writer studied the matter very closely while learning, both where he was employed and on the neighbouring farms, but never came to any satisfactory conclusions. Some farmers bullied and drove their men, doing them down to the last halfpenny in paying their wages; working on the old principle

'Tis the same with common natures,
Treat them kindly, they rebel.
But be rough as nutmeg graters—
And the rogues will serve you well.

This type of farmer got quite a lot of work done, as long as the men were watched, but of course never kept their labour very long. On the other hand some farmers treated their employees with every kindness and consideration which their means and the state of farming would allow, but this I noticed was almost invariably abused. If the men were late owing to bad weather and travelling conditions, and the farmer said nothing, you would not find them stopping a little later or coming earlier when conditions improved. No, they would come trailing in as late as they were when the roads were blocked with snow. If in the kindness of his heart the farmer sent them beer or tea out into the fields in hot weather, they accepted it as a right, and would hang about if it was delayed, instead of getting on with the job. In winter if they were provided with a warm room with a fire in which to eat their meals, they would not even keep it tidy. If a man was receiving milk, firewood, and a free cottage, he would grumble if he heard another man was receiving an extra couple of shillings a week, but without that payment in kind.

Either type of farmer, the bully or the Christian, would tell me his whole time was taken up in supervision, and this on three or four hundred acres employing six or eight men. Assuming the farmer's income to be £1 an acre, or £300 to £400 a year, then each labourer earning say thirty shillings a week in those days, required £1 worth of supervision weekly.

Other farmers, of course, devoted their attention to hunting, shooting, fishing, tennis and golf, balls, garden parties and cards, leaving their men to do as they liked; and who could blame their men for a 'bolshy' attitude, if the farmer was not attending to his business, a state of affairs which was reflected in the output of work, and sooner or later in the bankruptcy of the farmer?

There were, of course, a few good, conscientious men, who served their employers faithfully and well, and there were a few employers who abused their service. The best cowman I ever knew, who taught me much of value, had not missed a single milking for forty-three years, yet received ten shillings a week less than the other men on the farm, because his employer knew that the old man was afraid of being unemployed. Actually he could have got a job anywhere, and compared with some cowmen I have met would have been cheap labour at double his wages. An extreme case which labour leaders would seize upon with glee to press their case; although they should also explain away why a farmer of my acquaintance paid to a labourer's widow the wages which her husband would have earned, for seven years, to enable her to keep her family of five children together, while the trade union to which he had paid in for years did not refund a penny. Probably they would say that such is not the function of a trade union; the same could be said of the farmer who hires his labour by the day or week.

Summing it all up, it seemed that a farmer either had to drive his labour and lose workers frequently, or accept a very low output of work and tolerate much that was slack or careless to keep his men, and pay them as little as possible, for they could not justify good wages. We had got to have something very different at Oathill.

In this district farm work was regarded by many of the villagers as the last refuge of the inefficient and mentally defective, for the others could obtain work in the towns. I remember a lorry driver coming into the farm and, finding another who had skidded off a slippery stretch of ice into the ditch, saying in terms of deepest disgust, 'Why! You ought to be at plough'. As if ploughing were a far less skilled occupation than lorry driving!

A local boy, who had come to us straight from school, gave notice after a couple of years. Asked for his reason, he said he liked the work, was very fond of the animals, could not better his wage, *but the other village boys laughed at him for working on a farm.*

Not only has the shortage of labour become more acute as the years go by, but the standard of work has steadily deteriorated. True, Shakespeare said that servants were not so good as they used to be, and I am glad to say something in common with him in my literary efforts, but it can be measured by output of work and the skilled labour available. It is quite common in this district of big farms for a farmer to have only one rick builder and thatcher on the staff. We know one farm employing twenty men, not one of whom can carry sacks of wheat from the threshing machine to the granary; and another where a local smallholder has to go to do the sheep shearing. We ourselves have been 'phoned on a Sunday morning by a farmer who wanted to know if one of us could go to milk his cows. His two cowmen had met with an accident on the way to work, and neither the farmer nor his other eight men had ever milked a cow.

At the present time I do not know of a young, really good, all-round farm worker who can turn his hand to anything, in any of our local villages. There are a few skilful tractor drivers, but where will I see a well-turned-out team of horses? But it is in hand work that the loss of efficiency is most marked.

I have before me the recommended piecework rates for Oxfordshire, issued by the War Agricultural Executive Committee, agreed between representatives of the National Farmers' Union and the National Union of Agricultural Workers for 1943. I have also a wages book kept on a local farm over a long period, which shows most striking differences between day work and piecework. I take mangel-hoeing, thatching, and manure-spreading as typical examples.

### MANGEL-HOEING

| Year | Per day | Per acre |
|------|---------|----------|
| 1857 | 2/– | 4/– |
| 1913 | 2/6 | 7/6 |
| 1923 | 5/– | 15/– |
| 1943 | 10/– | 60/– |

### THATCHING

| Year | Per day | Per square |
|------|---------|------------|
| 1857 | 2/– | 6d. |
| 1913 | 2/6 | 8d. |
| 1923 | 5/– | 1/– |
| 1942 | 10/– | 4/6 |

### MANURE-SPREADING
20 loads per acre

| Year | Per day | Per acre |
|------|---------|----------|
| 1857 | 2/– | 1/– |
| 1913 | 2/6 | 1/3 |
| 1923 | 5/– | 2/6 |
| 1943 | 10/– | 12/– |

It will be seen that while the daily rate has increased to five times, the cost per acre for hoeing has increased to fifteen times, thatching per square to nine times, and manure-spreading to twelve times, which clearly illustrates the falling output of work per man on piecework against day pay.

Any really skilled worker could earn £1 a day at the rates at present fixed; Irish labourers have told me they have received as much as £2 a day on the agreed rates. This would indicate that the rates are too high, or the ordinary wage too low, or what is far more probable, that the output per man on daily work is too low, and he would not earn more calculated on a piecework rate.

I have always believed that everyone should be able to enjoy the fruits of his hard work, ability, or good fortune, and I do not grudge anyone their £1 a day at hoeing. I could earn it myself, while I have earned £1 an hour with my pen. But I also contend that those who can do the work should earn the money, and for that reason I consider that a member of the Women's Land Army hoeing carefully and thoroughly is unjustly paid at 9s. a day when an Irish labourer, from neutral Eire, should earn £2 even if skimping and scamping through his work. I have hoed against them stroke by stroke and you cannot make a first-class job when getting over the ground is the only consideration.

The same applies to Italian prisoners. I do not know how much they earn, but when five go to work on a local farm another devotes his attention to cooking them a hot meal midday, and does not do any work. The Land Girls have bread and jam sandwiches which they have had to cut themselves the night before. Somehow I feel that the vast powers of which War Agricultural Committee members boast might be utilized to obtain equality of treatment for those under their direction, either by equal pay for equal work, or equal conditions and amenities. I trust the digression will be pardoned; I use it to draw attention to the problem which we had to overcome in making the best use of labour and the land.

For our first year at Oathill we employed a good old-fashioned labourer; he was getting a little slow, but took a real interest in his work and served us well. Then for the next five years we did not employ any labour except a gang of casual workers for threshing; and how *casual* they were. A lot of sympathy was given to the unemployed in those days, and it was said that the greatest service one could render to the country was to provide employment, yet in our experience most of them were mentally and physically incapable of anything like a day's work. We used to watch them coming in the morning, and at a slight rise in the road leading to the farm they would flop off their bicycles and walk up the slope. If a man feels like that in the morning, what will he be like by night?

On one occasion the threshing-machine engine broke down on its way to us, and as the casuals had been engaged to start on a certain day, and we did not wish to turn them off, we set them to work ditching. They worked fairly well and had half finished by the time the thresher arrived. On the completion of the threshing they all lined up for their money, and although they would all be going back on the dole, not one asked if he could have the job of finishing the ditch.

When we had sufficient stock to need labour, we found the only way was to have boys from school and train them as we wanted them. They were quite easy to get, as few farmers will be bothered with boys of fourteen, whom they will lose into local industry after a year or two. We did our best to teach them, and they served us quite well. We also tried to interest them in a profit-sharing system, but in their lack of general education, and their inbred suspicion that farmers exist for, and by, doing down their employees, distrusted it and preferred to have a shilling, or two above the standard rate for the district, which they could spend by the next Friday night, rather than draw a substantial bonus at the end of the year.

While I was learning farming I made up my mind that when I became a farmer I would take pupils, not with the object of obtaining big premiums or cheap labour, but really to teach them their trade as I would like to have been taught; I also intended to remunerate them at the real value of their labour, so they could get a better start in farming than I had had. I believed then, as I do now, that the only thing wrong with British agriculture was the lack of really capable and progressive farmers, and well-trained workers. I believed too that the solution of nearly all the farmers' difficulties lay in their own brains and within the boundaries of the farms; and that far more could be achieved at home than in passing resolutions at the local meeting of the National Farmers' Union, designed to bamboozle the Government into bolstering up inefficient methods at the expense of the taxpayer. At the same time I felt that no-one is qualified to teach until he has proved his own theories. So we did not take a pupil until we had established ourselves as tenant farmers and then bought the farm freehold by our own efforts.

So carefully thinking out a fair system, we started with one boy. The difference was so striking between him and the local labour that for several years we have run the whole farm with pupils. Our normal staff is three. We set out to give them a really straight deal, and have been repaid a hundredfold by their loyal and wholehearted service. Most of the credit is due to my mother and sister, who keep them happy, comfortable, and well fed in the house. Outside, we teach them their job, pay them on the profit-sharing system, in which we tried to interest the local labour without success, and by which our pupils can earn sufficient in a few years to take a farm of their own, if necessary with financial assistance from us; for where could we find better investments than in backing those we have learned to know, and trust, and who have thoroughly mastered our foolproof methods of making money in farming?

Dr. C. S. Orwin, one of the great authorities on agriculture, says in his book, *Speed the Plough*, that 'farming is closed to all who would pursue it other than as labourers, unless they have money to invest, while in industry education and technical ability is sufficient to reach the heights'. Our pupils have proved otherwise. Personally, I believe there is more scope and less competition in farming, for a man with a well-stocked brain, than in the whole of British industry. Of the last twelve pupils we have had, three are farming, nine are in the Services, serving their country, until such time as they can take up farming.

How do we select them? An advertisement in the *Farmer and Stockbreeder* or *Farmers' Weekly* will bring in a good bunch of applications. Four or five who write the best letters are interviewed. No-one is accepted without interview and suitable applicants are given the opportunity to come for a month on trial. Sometimes, of course, a boy is recommended by an old pupil or business acquaintance, but influence counts for nothing, he can only be accepted on merit.

What are the qualities we look for? Average height, build, and weight for his age; that he has not suffered from any serious illness or accident, for those who are taking up farming for the sake of their health have not the drive and energy necessary for the job; having worked on a farm during holidays, or kept small animals or poultry as a hobby, is a good recommendation; manual dexterity, indicated by a fondness for woodwork, or interest in science or biology—for agriculture is an applied science, also help. An enthusiastic Boy Scout, other things being equal, is almost sure of the job, for there is much in the scout training which is invaluable on the farm. We have always regarded Lord Baden-Powell as the greatest educationist in the world. Sometimes, of course, an applicant has had experience elsewhere for a short period, in which case we like it to have been a very good or a very bad place. If the former he can carry straight on, if the latter he will see the contrast. While a good general education is desirable, academic standards count for nothing. Of six boys, three of whom had obtained School Certificate, and three had failed, the last proved most successful. This is remarkable in view of the requirements for most trades and professions, and the examination is really only a fair test of general knowledge, though it does tend to select precocious children who can trot out the right little answer as required, against those who learn slowly but never forget what they have been taught. I think the real reason is psychological; the certificated thinks what a wonderful fellow he is, and is disappointed that the farmer does not share the same opinion, while the unsuccessful boy has been told so often by his schoolmasters that he is utterly useless, that he is gratified to find that the farmer looks for very different qualities which he usually has in full measure. In the same way the boy who has come in conflict with school authority is sometimes a great success, for he finds in farm work an outlet for his physical energy; while the good little boy who tells you he has never had a fight in his life usually gets homesick, and wants to run home to his mother after three days.

To have been to a farm institute or agricultural college is a disability which few boys can overcome on this farm; either they look on agriculture from the detached academic point of view, or they are looking for the opportunity for fooling about, which we will not tolerate. Our pupils usually feel the same. I remember overhearing one saying to another who was here on trial, 'This job may be the only opportunity I shall ever have of becoming a farmer and I intend to make the best of it; if you want to play the fool go back to college where there are two hundred more like you'.

Sons of farmers, business, or professional men have done best, and for that reason are favourably considered, though we have no social or religious prejudices. Individual merit is the only consideration. With sufficient money anyone can boast that he was educated at Eton and Balliol; only a boy of character and ability can say he learned his farming at Oathill.

*LIQUID MANURE BEING MIXED WITH PEAT MOSS FOR TOP-DRESSING*

*PASTURE, 1943. STOCK GRAZING CAPACITY TREBLED BY PENNING WITH FOLD UNITS*

We do not always find all the good qualities in any one individual; one does not when buying a horse. Horse-buying seems to me far easier, for in twenty years of farming I have only sold one horse for less money than I gave for it, while I have often been mistaken in selecting a pupil.

A pupil lives as a member of our family, with free board, lodging, washing, and insurance stamps. He receives ten shillings a week for the first six months, rising to £1. In the second year he receives in addition a bonus based on the previous year's output. So that in pre-war days a boy of seventeen or eighteen in his second year received the equivalent of £3 a week, allowing £1 for board, lodging, etc., against the minimum agricultural wage of 35s. for a man. In wartime the bonus has been as high as £90 per annum. One boy told me he had saved £150, after keeping himself in clothes and travelling expenses in twenty months.

Not only do we pay good wages, but the learners actually earn them. A good boy after twelve months' training in our methods is far cheaper labour than any we can hire locally at half the money; without exception they all grow and put on weight, which would indicate that our high-speed methods have no deleterious effect on health.

Not only do we teach the practical work, but the scientific and economic aspects are carefully explained. It is quite common for a boy to acquire a better knowledge combining practice with theory in a couple of years, than by four years in an agricultural college. How much better he learns all the names of the weeds that grow in the fields, and the families to which they belong, if he is told them day by day while hoeing roots all summer; the points, bones, and organs of the animals as he grooms them each morning; meteorology, if he is asked his weather forecast for the day at breakfast-time.

I always tell them the labourer should know how, the farmer should know why; that there is a reason for everything we do on the farm, and it can be given, whether it is the way to litter down a loose-box, or the order in which a horse's harness is put on; that a penny-worth of thought is worth a pound's-worth of manual labour; that the correct way is the easiest in the long run; that one should visualize the whole, but concentrate on the details; that everything I teach is the recognized standard practice of the best farmers, and by following it they will be accepted as good farmers from Land's End to John o' Groats. For that has been my experience.

What a difference we find between these boys and the ordinary run of labour. How often, in the past, I have had a local boy doing something quite wrong and thought to myself, 'Now I don't want to upset this chap, but I must go and carefully explain where he is wrong'. Result? I get the reply, 'If I'm not doing it well enough for you, give me the money due to me, and I'll clear out'. How different with the better type of public or secondary schoolboy. He says, 'I'm sorry. How foolish of me', and he does the job as he is shown.

To some people, of course, we would seem very fussy and particular in the way we have things done. But good work is only a matter of habit. We have all been told the old proverb, 'Sow an act—reap a habit. Sow a habit—reap a character. Sow a character—reap a destiny.' I give just one example of care and attention to detail which is typical of our farming, and applicable to any other job on the land—and the results achieved by it.

Everyone is familiar with how sheaves of corn are set up in the field after cutting in shocks or stooks. When we do this each pair is set with the knots of the string outside, for the sheaves stand and drain off the rain better this way. Then they are pitched in pairs to the wagon, the man on the load taking one in each hand, opening the pair, and the knots are up, thereby making a compact and tidy load. At the rick, the sheaves are pitched the same way and built into the walls knot up, while in the roof knot down. In this way, with a suitable slope on each sheaf, you get a weatherproof rick. When you come to thresh, the knot comes into the hand of the man cutting strings, which are saved and can be used for other purposes, and thus is avoided the danger of stock being killed by eating string in the straw. Now many farmers will say there is no time for these refinements. Yet our harvest is invariably finished first in this district, and a threshing-machine proprietor has told me that we thresh cheaper per quarter of corn threshed than any other farmer he knows. Some people of course, buy dyed string, so that it can be picked out daily when the cattle are fed or littered, yet it is far better to keep it by means of a little efficiency in stooking, stacking, and threshing. Since the war I have seen many illustrations in the papers showing soldiers and land girls helping with the harvest. In not a single instance have they been pitching sheaves correctly. Quite recently I saw five photographs showing members of the Women's Land Army being specially trained as forewomen, these being all carefully selected, yet in three cases they were doing something which was demonstrably wrong.

All the pupils we try are not a success. We can only teach when people are interested. We cannot help the lazy or dishonest, for there are some who would be quite happy to let others earn their bonus for them, and accept a free hand-out. But nothing is too good for those who are looking for the thorough training which will enable them to become capable and successful farmers, and who will help us run our farm as it should be run. We share our knowledge freely, will give any assistance they require to gain more experience or take a farm of their own.

While we prefer boys, we have had the pleasure of training others. A really capable man, who has made a success of some other career and is taking up farming later in life, can master in months the basic principles which it takes a boy years to learn, although he never achieves the manual dexterity of youth. Ex-officers and ex-servicemen generally, who are the most deserving of any assistance we can give, are terribly handicapped against business men. They have never had to think for themselves, everything is laid down in the King's Regulations or Admiralty Instructions, if any new work has to be taken up they receive a special course of training lasting perhaps several weeks, and they find it very difficult to adapt themselves to farm work and its entirely different outlook on life, where apparently one has no leisure; but has to study the theory and science while learning the practical work. The three fatal 'S's'— Smoking, Swearing, Standing about—are as common in the services as they are on inefficient farms. To me they indicate lack of self-control, and betoken the man who cannot think or act without lighting a cigarette, the man who shows his irritation when things go wrong, and the man who cannot tell himself to get on with the job.

But under our system it is not all work. We allow one clear day a week, seven days' holiday first year, and a fortnight the second. These days may accumulate if desired and be taken as short holidays from time to time, which our pupils invariably prefer. With fifty-nine days the first year and sixty-six the second, they can fit in several little holidays at home, and any days they have not taken by the end of the year are compensated for by extra money. Generally speaking they take say a week every quarter, just to go home and see their parents, and cash in the rest, for they are usually saving hard for the day when they can take a farm of their own. Sundays are our equivalent of a half-holiday, for there are the animals to look after. We have never believed in Saturday half-holidays either for ourselves or anyone else. If they want to do any shopping, and are prepared to go between milking and feeding times, on days when no urgent field work is in hand, we do not count it as time off. Though it is our rule that anyone doing his full share is always given any time off he asks for, we know that he will not ask unless it is really necessary. For it must be remembered that a farmer can take time off, providing he is not neglecting his work, and our whole system is based on teaching each individual to think like a farmer. When this is achieved, he is more than half-way to being a farmer, for he has developed a sense of personal responsibility towards his work, and he does not have to be 'clocked-in' or supervised.

The holiday system is also valuable because it gives each a change of responsibility, inasmuch as he has to take charge of stock for someone else, and also leave his stock in someone else's care. By the end of the year any pupil can take charge of the daily routine, or instruct others in the care and management of the stock. On the farm we try and arrange that each shall do his share of the seasonal work, ploughing, hedge-laying, sheep-shearing, stacking, thatching, and the rest. With a fully equipped workshop many other useful things can also be learned, for a farmer can save a lot of money if he can build his own poultry houses, repair his implements, and execute the hundred and one little repairs which are so often neglected on the general farm. The accounts records, and books are also available, so that each pupil can learn how, when, why, and where the money is made or lost. The standard textbooks on agriculture and stockbreeding can also be studied.

Now this is not a prospectus, designed to obtain pupils. I have described our system in detail simply in the hopes that other capable and successful farmers will do their share in teaching the rising generation to carry on their good work. We have found it well worth while. While the farmer's first duty is to the land, his second is to share his knowledge freely, so that other land may be better farmed.

If my varied experience in learning my trade, experience in interviewing applicants, and teaching is of any value, perhaps I may be qualified to advise those who wish to take a post as a pupil. My advice is as follows. Learn on a small mixed farm, where the farmer is a real worker, and making a financial success of his business. Beware of the gentleman farmer, for if the man is not working himself you seldom get enough to eat. Only those who have worked all out between the age of sixteen and twenty know how much food a growing boy requires. Beware of the man who is chairman of lots of committees, for he will be far too busy to teach you anything. Do not accept a situation to lodge with a farm worker, for you will begin to think like a labourer, but only where you can live with the farmer, for you must learn to think like a farmer. Be prepared to take responsibility, if it is only seeing that you keep your time in the mornings; whoever else oversleeps you should be up and doing. Do not bother how hard or how long you work, as long as you are learning. It isn't the smooth or the easy which will make you a capable farmer, so never dodge the unpleasant task, just master it. Make it a rule to do just a little more than you are expected to do. Keep a diary and notebook, for there is so much in farming which you cannot carry, in your head. My notes on farming in the four years I was learning ran to a hundred thousand words, so you should be able to jot down something interesting and useful each day on therfarm. It is a good plan too to put a plus or a minus according to whethe you were praised or blamed during the day, remembering that there is a reproach in unmerited praise, and it counts against you if that horse bumped his foot against the gate when you were taking him through—even if the boss did not see it. If you do this, and are really giving your whole mind to the job, you will see how the plus marks increase and the minus decreasse. To become a farmer within our years by your own efforts is posible, both I and my pupils have proved it, but to do it you have to fill the unforgiving minute every lay. But believe me, it's well worth while—so good luck and good farming.

# CHAPTER NINE

## CORN BINS UNLIMITED

Many people may say, 'It's all very well to pose as farmers, but are you not the manufacturers of the corn bins that you see on the farms all over the country?' This is quite true, but as it has grown out of a sideline in farming it may therefore be of sufficient interest to be included in this book.

At the bottom of the depression in agriculture our only means of making a profit on growing grain was by cashing it through stock. In those days it was possible to walk round the boundaries of our farm and looking over the hedges to see every adjoining field tumbled down to grass; yet when wheat dropped to eighteen shillings a quarter, or a third of the cost of production, we could still keep our land under the plough, by converting the grain into fattened poultry, for five pounds of meal will make a pound of chicken, and you could sell two-pence-halfpenny worth of wheat for a shilling in a different form. How thankful we have been since that it was so, that however bad the times we could go steadily on, for above all a farmer does need faith in the land, and to stop the plough brings in its train a whole cycle of misfortune and ruin.

To feed grain or meal daily in large quantities entails efficient storage, for the meal must be kept round from harvest to harvest. It must be protected from rats, mice, and damp. To tip it down in a barn encourages vermin, while damp destroys its food value and mould will spread diseases like aspergillosis and laryngo-tracheitis. So here was a riew problem for us to solve.

We would need a dozen large corn bins, which are large rectangular steel boxes capable of holding a ton of grain each. These would cost £6 to £12 apiece, which was far too much capital to tie up; so we had to design and make our own, which we found we could do for half the list price. They seemed to be quite satisfactory in every way, did not bulge or twist, and the grain stored perfectly.

Then one winter evening, while walking to the village to buy some stamps, we were discussing the bins and one partner suddenly said to the other, 'Why not advertise in the local paper, some other farmer may need them too? It's very improbable they can make them themselves, and if we get an order or two we can make the bins after dark when we have finished our farm work.' In the remaining mile and a half the advertisement was composed, a postal order purchased, and on returning home the letter was written, and a note made—but not in the farm books—'Advertisement, 5s. 4d.'

The advertisement appeared on the Friday and on Saturday there was an order for a bin at £2 14s. to hold a ton of wheat, carriage paid. It was only a short distance we had to deliver it, and as we spent two evenings on its construction the farmer received it by the following Thursday. We received cash on delivery, and on working out our costs found we had received fifteen shillings for our labour. At this stage we decided to launch out on 'big business', and we risked the whole fifteen shillings on an advertisement in the *Farmer and Stockbreeder*. This advertisement also sold a bin, and we again risked our profit, with the result that the next appearance sold two bins. We now had our capital expenditure returned and a small balance in hand, so it will be seen that the whole business was founded on 5s. 4d., the materials being bought on a month's credit from a local ironmonger and timber merchant who gave us the customary month's credit; and as we were selling on a cash basis, we had the money before we had to settle the accounts.

In the first year we sold fifty-two bins, earning about a shilling an hour. The agricultural depression continued, unemployment figures in the local villages continued to rise, and we wondered if we could make a new rural industry, for which there was obviously a pressing need. We tried one or two local men, making the bins in our cartshed, but they would only work if one of us was there with them. If bins had to be delivered to the station, or other work attended to, they did not justify their wages. Gratitude is the rarest of human virtues and they could not show their appreciation of being found a regular job by a steady output, although they had families to support, and we found at the end of three months that we had barely cleared expenses, for we now had a regular advertising charge to meet. We changed our men, with the same result. It seemed a great pity, for we could earn good money making bins in the very limited spare time we had, wet weather and winter evenings, and would have liked to share our good fortune with others, and we would have been prepared to work on a very small margin of profit to cover the general expenses of running the business. We have since learned that there are a great many people in this world who are not worth helping, and you can only help those who will help you. I once read how a famous naturalist used to light a fire in the forest, and all the monkeys used to come and play round it and thoroughly enjoy the warmth; but never by any chance did a monkey pick up an odd piece of firewood and keep the fire going. Is evolution such a long process when you notice the principle on which some people work?

So we dispensed with the labour—but we were destined to carry on the business. The stopping of our advertisements caused a flood of orders, from farmers who had seen our offer week by week and rushed to order before they were too late. We felt we could not return the money, so worked harder than ever to turn out as many as possible by our own efforts. I started the farm work at three-thirty in the morning and carried on to 7 o'clock at night, and then went and did two hours' work on the bins or on correspondence and accounts in connection with bins, so that my brother could devote all his time to bin-making, for he was always a much faster worker with wood and metal tools than I.

The orders and the money rolled in, for by this time we were getting repeat orders from our earlier customers. Then one day a boy of fifteen walked in and asked for a job. We set him on, and he proved a steady and conscientious worker, with as good an output as the men we had tried to teach. Here was the solution, as on the farm, of our labour difficulties: good boys, carefully selected, and trained in our own methods. After a couple of years the business outgrew our cartshed and we built a workshop, our staff increased to ten, so here we had a rural industry, providing useful employment and training skilled carpenters and tinsmiths, without any help or subsidy of any sort, and supplying the farmers with a first-class article at half the usual price, for our standard bin was now listed at fifty shillings, against an inferior article at £ 5

12s. by our nearest competitor. The time of making a bin had been reduced from seven man-hours to forty minutes.

Only local labour was employed, boys recommended by the headmaster of a local central school coming straight from school at fourteen. The senior boy of our staff was always the foreman, unless as sometimes happened he did not care to take the responsibility, though it was worth ten shillings a week extra, when the next one would be given the chance. They were usually skilled workers after two or three years, and stopped with us six or seven on an average. How strange it was that these boys would take pride and interest in hammering galvanized steel all day long in the factory, looking down on farm work with its variety and interest as something performed by a lesser mortal, so deeply set is the corroding influence of the industrial revolution biting into rural England. For judged by ordinary standards these boys were all of a superior type and intelligence. They welcomed the profit-sharing system, and there is nothing like it for running a trouble-free factory. Wages were a certain precentage of gross sales. Each member of the staff drew a basic wage, and the difference between that and the percentage of the value of the goods sold, in the form of a bonus at the end of the month. Unlike piecework, when each employee turns out as many components as possible, but has to wait if the man below him on the production line is hindered, the profit-sharer goes back to give him a hand. For each started his training on the first stage of construction and worked up along the line, and is therefore expert at every process by the time he has reached the top and can pass out the finished article. There is no shoddy work (just good enough to pass the foreman), for each one on the line is expert at the process before him and will not accept it; in this way no material is wasted, and each one knows that poor work might involve a lost sale and that means a smaller bonus.

Once started the credit of running and building this business was entirely due to my brother, Frank Henderson; the formation of a limited liability company to keep this branch of the business separate from our farming was to him a natural development. To this day, as chairman of the board of directors, I feel a complete fraud, and in signing documents which require my occupation to be stated always put 'Farmer', and never 'Company Director'. In fact, I never thought of myself as such, until one day our accountant, coming in and finding me unloading a couple of tons of coal, remarked it was the first time he had seen a company director so engaged. *Sic transit gloria mundi!* Nevertheless there is now a Henderson Corn Bin in nearly every parish in Britain, and in a good many places abroad. A famous statesman once said that 'exports are not determined by a trade balance of millions of pounds, but by individual transactions of five pounds', in which case we have had our share, and contributed to the industrial might of our country, with an original capital of 5s. 4d., and a desire to provide employment in our own locality.

Many people have expressed surprise at the low prices of our goods. In fact a number have called to buy a bin and said that they had had one before from us, and thinking to save money by making their own had ordered the materials, to find that they cost more than the finished article from us carriage paid to their nearest station. The explanation is that as the business grew we purchased materials in very large quantities, sometimes wholesaling them, at such low rates that we could reduce our prices accordingly. At one time our 50s. bin was sold for 37s. 6d. It caused us a great deal of thought in the early days before we risked buying ten hundredweight of steel sheets; latterly I sometimes saw orders in the office for two hundred tons of steel, nails in eight-ton lots, hinges by the hundred gross; but my partner had not thought them of sufficient interest to mention. Two hundred thousand feet of timber was quite usual, to ensure the right quality. It was all self-sown Russian pine, at least seventy years old and seventy feet high, with a breaking strain greater than the finest English oak, and would not split, twist, or warp. Machinery was installed from time to time to reduce hand labour, including several devices designed for a special purpose by my brother.

It is interesting to note that the method of construction has never been altered, simply because we have never been able to improve upon it. The number of types and sizes has increased to thirty-two. In addition we manufactured for a number of other firms who sold similar bins under their own names, and it was our proud claim for some years that every order was executed and on rail within twenty-four hours. This even applied to special sizes to fit into our customers' buildings. This necessitated some very careful planning; for example, a bin was required on a Shropshire farm to hold six tons of oats, but owing to the position of the doorways it had to pass through a window two feet by two and a half. This meant making over a hundred sections each one foot eleven inches by two feet five inches, but it was on a lorry within twenty-two hours from the receipt of the order, and filled with oats within forty-eight hours.

With the organization of the corn bin business at our disposal many other lines were taken up and developed, but we have never permitted it to interfere with our farming. Once it was established my brother spent a lot of his time on the farm, though within call if required. In the difficult farming years, people would often say, 'I wonder you bother with farming, when you have so many other interests'. Yet to our way of thinking, who could compare even business-building with serving the land? We are farmers at heart and never had any doubts where we prefer to be, but also our business experience showed that there is nothing which pays like a well-stocked and managed farm. There is a lot to be said for having your cake and eating it, even if the old proverb says that it cannot be done. It has also always been our rule to keep the businesses quite separate; money made in farming is re-invested in the land, that which is acquired in business is reinvested in business.

Great difficulties were experienced in the early days of the war as no Government department thought that corn bins were essential articles on the farm, although we knew that well over 20,000 tons of meal and grain were stored away in them every night of the year to keep it safe from rats, mice, and damp. We experienced the very greatest difficulty in obtaining permits for materials—and finally we gave up the battle against departments who had never heard of a corn bin and did not know it was necessary to store food for animals. At a modest estimate 5,000 tons of food have been destroyed and wasted on the farms of Britain for want of the bins we could have supplied.

However we hope to start again after the war, for we owe a duty to our staff, now serving with the forces, to provide them with employment in the days to come.

# CHAPTER TEN

## HOLIDAYS

The school of thought which believes that farmers should not have holidays may skip this chapter. To them we tender our apologies, and trust they find excellent value for their money in the rest of the book.

In the first five and a half years we did not miss a single milking, in fact only on three occasions did we take a few hours off during all that time. Working from 4.30 a.m. to dark, and every penny we could find reinvested in stock and on the land, we had neither the time nor the money for amusements. Our only recreation, apart from reading (for our friends were very good in lending us books), was the study on Sunday afternoons of the geology, botany, archaeology, and kindred subjects, within walking distance of the farm. It might surprise many people to know of the absorbing items of interest which can be found in such a small area, which is barely mentioned even in the best guide books.

We found a reason and purpose for everything. The origin of the name of this farm mentioned in the first chapter is an example of this. The next farm, Whitehouse, originally Whiteway, is situated on the old straight track from Colchester to Pembroke, without a deviation which a surveyor could detect, never suspected by the leading archaeologists, but clearly marked in our area by standing stones, including the Hawk Stone through which one may sight on the line. To Mr. S. Watkins, the author of *The Old Straight Track*, must be given the credit for suggesting that these tracks from three to four thousand years old existed. But we had the thrill of finding this one, probably the finest in the country, when we climbed the Hawk Stone and gazed through the aperture. For to detect it you must have, like the wheat inspectors already mentioned, insight as well as eyesight. On this line travelled the early prospectors in search of metal, and traded salt, hence the Whiteway farms, which to this day are only found on the line. Across the whole breadth of England and Wales one could travel on foot, missing the swamps and rivers, for the track travels along the watersheds, and crosses the Severn in the only place where it could be forded. Hundreds of the sighting stones have gone, on their sites tumulus, church, and cross-road have been made, but the track remains clearly marked for those that have the clue to detect it. Personally we doubt if it was built for trade; that must have come much later; it probably had a religious significance of a phallic nature for the early inhabitants of these islands, and the stones were preserved by a superstitious awe instilled by the early priesthood, and passed down by generation to generation long after their origin and purpose had been forgotten. It is interesting also that the line is parallel with a modern degree of latitude, and could be set with a simple triangular device sighting on the old North Star, in the constellation of Draco, for this was he star of the Neolithic shepherds, and it is only in the last two thousand years that the Pole Star has come into general use. That period of course is but a short time, to those of us who for a brief span have minded the flocks on the hills, and gazed at the stars.

In this respect also the Rollright Stones, which are similar to Stonehenge, just a few miles away, have made a very great appeal to me. Perhaps I should not venture to express an opinion when so many eminent men have written whole books on the subject, but they respond to the calendrical theory, passed on to me by the old Gallowayan shepherd from his Celtic, ancestors. The ignorant Saxon and Dane invented fantastic nonsense about them, which local people repeat to this day, yet their message is there to read for those who would know the season of the year or the age of the moon on any given date, and that after three thousand six hundred years. The exact date of their erection could be worked out, only they appear to have been reset at random and only eleven stones appear to be in their original place; though should any farmer lose his gestation table he could still determine from them when his mare, cow, or ewe will bring forth her young. This to me is far more interesting than any nonsense about Mother Shipton, or the stones rushing down to the brook for a drink on Midsummer Eve.

On a three-hour walk by footpath from this farm, thirty-five of the thirty-six native trees of Britain can be identified. The rare plants we have found growing in waste places I shall not mention; we have been content to find them and leave them, giving no clue for the ardent collector to lay waste the countryside.

We could have taken the eggs of a hundred species of birds, yet not one has been disturbed. We believe they maintain the balance of nature. As farmers we never destroy a oaok, which in our opinion does far more good than harm; like poultry there are times when they must be kept off crops, but can render a valuable service to agriculture. Those who shoot rooks do so as an old English custom of the sporting farmer; if they were determined to destroy that which is inimical to their interests there would be no rats left in the countryside. We could trap a hundred young rooks a day in our poultry fold units by leaving the lids open at night with a little maize thrown in, but we never do. Even to a wayward fox which took to taking our poultry, and necessitated sitting up every night for three weeks before he came in range, I had to say, 'I'm sorry, old man, but——' before pulling the trigger.

However, I digress. With so much of interest in our own locality, perhaps we should not have ventured further afield, but as soon as we had comfortably established ourselves as tenant farmers, and could see our way to buying the freehold in a few years, we began to plan annual holidays.

We had often talked to school teachers and others, who seemed to make so little use of their long holidays, that we determined that the little time which we would have at our disposal should be used in such a way that a holiday would be something well worth remembering for the rest of our lives. Our muscles hardened in regular farm work could find rest in travel, which to some people is in itself exhausting, while our brains could 'regulate imagination by reality' as a famous traveller once remarked is the proper purpose of travel.

For myself I planned to see Britain first. I would travel through every county in England, Ireland, Scotland, and Wales, climbing the highest mountains, visiting the sources of the great rivers, seeing as much as possible of the countryside and of the farming, for a farmer even on holiday can seldom forget his calling. It took all my holidays for ten years to carry out the plan, but in every one of the hundred days I saw and did something really worth while, which a line in a newspaper, or an odd thought will bring back as a cherished memory. For a holiday is in three parts planning, doing, and thinking about it afterwards. I am not yet quite sure which I enjoy most.

My mother would laugh at us, saying that as soon as Christmas was past out would come the guide books and time-tables to be studied in odd moments, and how was that last five minutes on the third day going to be used to the best advantage? While we felt that this was a pardonable exaggeration, we did like to have a definite plan with an alternative for each day.

Looking back through my diary, I see that I have motored as much as 2,000 miles in a week, with two companions to share the driving. For unlike most motorists I prefer to be driven rather than drive, for then I can look at the countryside and not bother about the road ahead. If my driver takes a corner too fast and we go through the hedge, I do not bother unduly, for as I tell him, I have often been out with a horse which has taken charge and done the same thing, so why worry about a donkey? With so much to see on either side I have no time for pressing imaginary foot-brakes, for that is how most motorists spend their time when being driven. Farmers too have another advantage on a holiday of this description inasmuch as they are used to getting up at 5 o'clock in the morning, so that they can travel 100 to 150 miles on clear roads at the best time of the day, by the time the ordinary holiday-maker is thinking about having breakfast. Then having spent the day sightseeing, visiting farms, or climbing mountains, another three or four hours' driving in the evening will complete the 300 miles for the day.

Although I enjoy motoring, I have also spent many happy holidays on my flat feet. Travelling by night to Scotland or Ireland I could make the very best of my time and opportunities, for the finest of British scenery can only be seen on foot, and a car often becomes a hindrance unless you have someone to drive it round the other side of a mountain range, when you want to walk right over the top. Unlike the regular hikers I travel light, for though I can carry twice my weight in the form of an eighteen-stone sack of wheat when at work, on holidays in the hills I depend on Christopher North's philosophy:

Among the hills a hundred homes have I;
My table in the wilderness is spread;
In these lone spots one honest smile can buy
Plain fare, warm welcome, and a rushy bed.

While I cannot vouch for the honesty of my smile, I have yet to find the shepherd who would refuse me a night's lodging in the hills. For did not my old friend predict 'in the hills my spirit will guide and comfort you'? Even in the inhospitable regions of Scotland where sporting interests have closed the inns and depopulated the glens, and where the few employees are forbidden to have relations come to visit them without obtaining permission from the estate office, for fear the game should be disturbed, I have been made welcome. I remember once knocking at a lone cottage; the young woman was reluctant to take me in, when an old man sitting by the turf fire said in Gaelic, 'For the love of God, Mary, take him in. He is one of ourselves.'

I had an even more remarkable adventure in Ireland on Whit-Monday 1937. I was stopping for two nights at Killarney. And with a day that dawned clear and fine I decided to climb the three highest mountains in the Emerald Isle—Carrantouhill, Caher, and Benkeragh, all lying in the range known as the Magillicuddy's Reeks, and totalling some 9,400 feet. So hiring a bicycle I rode out to the far-famed Gap of Dunloe, a wild gorge through the mountains, a popular tourist route, where I found a group of men with ponies which they hire out for people to ride through the glen to complete a circuit of the district, by coach, horseback, and boat. They were most anxious to hire me a pony by fair means or foul. But I told them I had come to climb the mountains. At this they nearly wept, for was not each one individually the official guide to the mountain? But on a bank holiday they had to take people through the glen, which was obviously more profitable! Could I not ride through the glen to-day—and visit. Carrantouhill another day? I could not, but on holiday I am prepared to spend freely, so I said I would hire a pony, ride through the glen, and climb from the other side, if I could also hire a trustworthy boy to take the bicycle right round the range by road and leave it at a cottage marked on my map near Lough Acoose. It was no sooner said than done. The chief brigand, for these people live by plundering the willing tourists, said he would reduce the hire of the pony to four shillings, if I would ride it quickly through the glen, so that the men coming from the other end could bring it back in time for another journey. For, said he, 'I can see your honour is a horseman.' He probably would say the same to a potbellied stockbroker, but the blarney of these men is unique; for generations their only source of income has been the tourist traffic.

And so in a few minutes I was trotting his fastest pony through the gorge which divides the Reeks from the Tomies and the Purple Mountains. Comparable to Glencoe in Scotland or Glen Sligachan in Skye, only on a smaller scale, the rough track winds and twists through the valley, the brawling river intensifying the desolation and rugged grandeur of this famous beauty spot. The Old Red Sandstone combines with the limestone to litter the glen with multi-coloured boulders, while the aspiring summits throw intense shadows across the glen to contrast with the chromatic beauties of the grass- and heather-covered slopes.

By ten o'clock I had handed over the pony and was climbing the steep craggy sides on the southern slope of the range. The usual route taken by the guides from the north is easier, but I was glad to have seized the opportunity of also seeing the Gap of Dunloe, and the fine view of the Lakes, which one would not see from the northern ascent.

*THE SILO*

*THE NEW HOUSE IN COTSWOLD STYLE*

The view from the top was superb. Right out on the Atlantic a storm was passing north-east, probably a hundred miles away. On one side was the Kenmare River opening out to the sea, on the other Dingle Bay and Galway beyond. I could trace the Shannon to Limerick. A range sixty miles to the east was probably the Galtymore or Knockmealdown Mountains, for my map did not reach so far. Almost due south was Mizen Head, which I judged to be a similar distance.

Far below me in the Black Valley the shadows still lingered and I calculated that for four months in winter the inhabitants would not see the sun, for the smoke rising from the pinprick white houses showed me there were little crofts inhabited all down the narrow glen.

Normally I am never in a hurry to leave a mountain top, unless bad weather is approaching, but to-day I felt an urge to get on, and by three o'clock I had stood upon the three summits and had started to descend the western slopes, having traversed the whole range of the Reeks from east to west in five hours.

As I approached the valley I saw that there were several cottages at which my bicycle might have been left by the boy who had agreed to ride it round the hills for me. Then I noticed a girl standing at a gateway watching me, so I turned and went across to her. She looked at me keenly, greeted me with the single Irish, or as I know it, Gaelic word, 'Thig', meaning 'Come', for some ten thousand words in Irish, Gaelic, and Welsh are said to be the same or similar. I followed her, and she opened a door, and I expected to find the bicycle inside; instead I saw an old Kerry cow in the last stage of exhaustion stretched out upon the floor. I looked at the girl, she gestured towards the animal, and walked out. As my eyes became accustomed to the gloom I saw the cow was calving with some complication. So taking off my coat, for there was not a moment to lose, I found quite a simple mis-presentation, the calf being still alive, and in a few minutes I delivered it safely. Putting it in front of its mother, when she quickly recovered as only cows can, and started licking it, I knew that all was well, and went along to the house with the glad tidings. Would I have some tea, she asked.

The kettle was put upon the turf fire. I knew from experience that it would take three-quarters of an hour to boil, for peat heats slowly, so I settled myself down for a chat.

The girl was about eighteen, very pretty in an ethereal way, and of pure Celtic type, but showing signs of tuberculosis, with a cough, high colour, and skin stretched tightly over her collar-bones, for she was far too thin. She told me her parents were away burying a relation, for eighty people had died in the Black Valley since the last winter. They caught cold and were gone in a few weeks. The cow had been calving for seven hours; she could do nothing, but with the simple faith of a Catholic had prayed that someone should be sent who could deal with it. As she prayed she had had a vision of St. Ninian, who assured her that help would come from the hills. So she stood at the gate and waited for me to come.

'Who is St. Ninian?' I inquired, for I knew nothing of the Irish Catholic saints.

'An Irish monk at the time of St. Patrick, the patron saint of shepherds, who carried Christianity to Scotland in the fifth century, a hundred years before St. Columba, landing at Whithorn in Galloway.' At this I sat up, for this was the native village of my Gallovidian shepherd.

What was he like? I knew the answer before it came; a tall, thin man, with piercing eyes, and a long divided beard.

What a fine example this would be for an investigator of psychic phenomena, supraliminal communication; the fact that while I received an urge to hurry from the hill, when all my mental faculties had been refreshed, the girl in her intensity of prayer conjured up a vision of the man who had taught me, in the days before she was born. For how else can it be explained? You would have to wait a long time at the foot of the Magillicuddy Reeks for an ordinary hiker who had calved five hundred heifers in his time.

So in due course I found my bicycle half a mile further down the valley, and cycled the thirty miles back to Killarney, visiting the far-famed Torc waterfall, and other wonders of nature on the way.

I mention this incident to show how close that which is hidden touches the lives of us who live close to nature. The man who lives in towns and gains his experience of life from the cinema and newspapers, and spends his holidays at Brighton or Blackpool, will be sceptical about the story. Those who have lived in the great open spaces, perhaps also the very old and the very young, will believe it.

Looking back through my diary I can see that it would be possible to write a whole book on a single holiday, so I content myself here by briefly describing how much can be packed into ten days, so that I could return refreshed in mind and body for another year's work.

I worked till 6 o'clock on a Monday evening; and then caught the 7.40 train from Banbury, travelling all night to arrive at Glasgow in the early morning, in time to take a walk round the city and then join the 7.11 a.m. steamer down the Clyde. Probably this is the most interesting river in Britain, with its great shipbuilding yards gradually yielding place to magnificent scenery. On through the lovely Kyles of Bute to Ardrishaig on Loch Fyne by 1 o'clock. The forty-odd miles by motor-coach through green hills and woods to Oban were accomplished by mid-afternoon, in time for an hour's sleep, tea, and then a visit to a famous herd of Highland cattle in the vicinity.

The next morning there was time for a swim before leaving by steamer at 8.45 for Fort William and through the Caledonian Canal to Inverness. An all-day trip this, during which I made the acquaintance of a Wiltshire farmer, also one from New Zealand and another from Australia. How restful for a farmer to sit in the sunlight on the deck of that wonderful seventy-year-old steamer the *Gondolier*, watching the reflections of the mountains slide past in the water, and chatting to congenial company. The evening was spent in renewing an old friendship and planning a trip for the following day.

The finest walk in Scotland, some with very wide experience say Europe, is right over the watershed from Loch Duich to Glen Affric. This twenty miles of incomparable beauty lies beyond the Great Glen which divides the Northern Highlands and the Grampian Mountains. Part of its charm may be due to its inaccessibility; it can only be attained by having two cars, one party going 120 miles to the west coast, and the other 40 to Glen Affric, if the centre is Inverness. Then by taking a 16-mile walk and changing cars it can be completed in a day. Once started it must be completed for there is no place of shelter, as the country has been rendered as inhospitable as possible by deerstalking interests, so much so that over great stretches no path is visible, only rocks and tough heather.

So very early on a glorious summer morning I set out with three companions en route for Croe Bridge at the head of Loch Duich for the west to east crossing, while our friends started a little later to travel a hundred miles less to the end of the road in Glen Affric.

By 10 o'clock we had parked the car and were well on our way up Glen Lichd beside its brawling stream between Sgurr Fhuaran and Beinn Fhada, both towering over 3,000 feet above us. My pen cannot do justice to the scenery, which combines everything which is lovely in nature. In many parts a deep Alpine valley, enclosed by precipitous hills, sometimes widening to three miles or more, but nowhere bleak or barren, for the landscape is relieved by woods of unsurpassed loveliness, varied in character, graceful in form, and luxuriant in foliage. Great rocks crop out among the heather and trees. Herds of deer graze unafraid upon the slopes. Above all, in the northern sunlight shone the great amethyst-coloured peak of Mam Soul.

At midday we bathed in a tree-fringed lochan, and as we ate our lunch feasted our eyes on the carmine-splashed mountains, for the bell-heather was at its best, stretching down to a carpet of cranberry, bog myrtle, and juniper, from which rose sturdy old Scots pine, remnants of the old forest of Caledonia, with clusters of silver birch against the clear waters of the little lake. Here we should have met our friends, but did not do so, and went on to meet them less than a mile from their starting point in the late afternoon. Their leader had been let down by an old football injury to his knee, and could not risk sixteen miles of hard going. His companions hid their disappointment as well as they could that they, on such a lovely day, should miss the walk of walks, as neither had had sufficient experience of the hills to find his way, without their leader, even to come on and meet us. There was a solution. I offered to take them back through the upper glen, which was a little shorter route, between Beinn Fhada and Sgurr nan Ceathreamhnan, picking up the car and taking them back to Inverness. They were fresh, having rested during the day, and so back over the mountains in the long summer evening and the northern twilight we went. I enjoyed every mile, the pleasure of my companions, and the thought that few people can ever have made the double journey in a day. In the twelve hours we were in the hills we did not meet a single human being.

We were very late back at Inverness, having travelled 240 miles by car, and in my case thirty miles on foot, but it seemed worth while, for two who made that journey will never again see the exquisite combination of water, woodland, mountain, cloud-topped peaks, and foaming cataracts in the old, unspoiled forest of Caledonia.

At 10 o'clock the next morning I left Inverness by plane for Shetland, arriving at midday. The afternoon I spent fishing in a friend's yacht, went to a midnight picnic, for Shetland is the land-of-the-never-dark in summer. The following day was devoted to sight-seeing, travelling the length of the islands, which extend for over a hundred miles. Then the following morning I took a plane back to Aberdeen, or more correctly Dyce, which is the airport for the granite city. Then by train and bus to Braemar, in time to climb Morrone in the evening, which rises to 2,819 feet behind the village. From the top I gazed across the Cairngorm Mountains, for on the morrow I intended to scale their rugged fastness.

Making an early start, for I had had the forethought to hire a bicycle the night before, I rode out to Derry Lodge, ten miles from Braemar, and the beginning of the Larig Ghru Pass, the highest in Scotland, rising to 2,700 feet. Here I left the machine under the bridge leading to the stalker's lodge, and was ready to start on foot. Half a mile up the glen I found a hiker, just setting out for the day from his tent. He told me that he spent a week every summer camping there and knew the whole district well, and on that day intended to scale all the Cairngorm peaks of over 4,000 feet, nine in all, there being only eleven in the British Isles. I asked if he would care for company. He looked at my sports coat and flannel trousers with disapproval, and said that I did not look like a proper hill man. I humbly admitted that I was not, looking at his full regalia, but perhaps I could come a little way with him, and learn a little from such an expert mountaineer? He relented, and off we set up the slopes of Ben Macdhui, the second-highest mountain in Britain. At the summit I stopped to admire the view, which is very fine, stretching from Atlantic to North Sea, from Ben Laeghal in Sutherland to the Cheviot in the Borders. My friend had not time for panorama, for he had minutes to cut off his record, so on he went down the other side to Loch Avon. When I was ready I ran down the slope, for was there not a time when I prided myself on catching a mountain sheep on the steepest hillside? I caught him up in half an hour. He seemed surprised to see me again, and in due course we plodded up the steeper side of Cairn Gorm, 4,084 feet, the mountain from which the range takes its name. From there we passed in the course of an hour to Cairn Lochan (3,983 feet). Here we saw a thick wall of cloud creeping up the mountainside. At this the professional showed some concern, but I had already taken out my compass, without which I never venture into the hills, and taken my bearings.

From Cairn Lochan I steered him down to the Larig Ghru, for on each side the slopes fell away, and up the cliffs and corries swirled the driving cloud, giving a visibility of only a few yards.

Once in the Larig, the clouds lifted, and the slopes of Braeriach being swept clear by the wind, we saw far ahead of us tiny figures toiling up the mountainside. We used them as our pacemakers, for they were fresh from the hostel at Aviemore, and overhauled them before they reached the top.

From here we could travel along the backbone of the range for over two hours, seldom dropping below the 4,000-foot line, and passing half a dozen named peaks. His enjoyment of the glorious summer day was somewhat spoiled by the cleggs and flies which assailed my companion's bare skin, in the conventional outfit of the hiker. I appeared more suitably attired, and in common with all countrymen my tanned face, neck, and arms had no attraction for the insects.

So on we went, past the source of the Dee, the highest river in Britain, which rises in a glacier for most of the year, before taking its first six-hundred-foot leap from the 4,000-foot level. All the rare Alpine plants were in flower, a sight one can seldom see in this country. I glanced from side to side at the country spread for a hundred miles on either side, and at the wonderful cloud effects piling up far to the east, which portended thunder later.

On reaching Cairn Toul, an anxious consultation of his watch, compared with mine, showed we had lost seventeen of the minutes gained on the first slope of Braeriach, a disastrous state of affairs to an ardent mountaineer, so I offered to race to the bottom. It proved the roughest, most boulder-ridden slope I have ever traversed, with an occasional sand slide, but we reached the bottom in under half an hour, enabling us to cross the Dee and reach my friend's tent with thirteen minutes to spare. This enabled him to feel that he had had a good day after all, in spite of having to cart round an amateur, and he was kind enough to say that with sufficient practice I would become quite useful on the hills. He was a clerk from a jute factory in Dundee, and lived for his annual week in the Cairngorms, and so I found it in my heart to forgive him, for I too had enjoyed the day. So retrieving the bicycle I rode back to Braemar through the thunderstorm I had seen gathering from the heights of the Angel's Peak (Sgor an Lochain Uaine, 4,095 feet).

The next day I took the early morning bus to Blairgowrie, via the Spital of Glenshee, a distance of thirty-five miles, and probably the finest bus route in the Highlands, the road climbing to 2,200 feet at the Cairnwell Pass and negotiating the famous Devil's Elbow, a dangerous double turn, before gently running downhill through very impressive scenery. From Blairgowrie I went by train to Edinburgh, arriving in time to have lunch and at 2.15 p.m. going on an afternoon tour by motor-coach to Abbotsford, Melrose, and Dryburgh Abbeys, arriving back at 9.30 in time to catch the train for the south, so that I could be back at work the following morning.

Ten very full and enjoyable days were these, and typical in many respects of the hundred I devoted to seeing Britain first.

What great contrasts I found in such a small compass! On the Romney Marsh fifty sheep fattening to the acre, the most heavily stocked sheep pasture in the world. On a remote island in the Hebrides fifty acres of grazing (!) to one sheep, they stood waiting for the tide to ebb and maintained life by eating seaweed. They are said to be fatter in winter than in summer, for there is more weed available then. In the Shetlands I saw sheep sheltering from wind and rain in enlarged rabbit holes: incredible to the English farmer, but a commonplace to the Shetlander.

There is good farming too in the far north. Many a local farmer in the Cotswolds might well envy the crops grown on the deep Old Red Sandstone of the Orkneys. There only the wind is the farmer's enemy, the soil and climate comparing favourably with those of any part of Britain.

A Jersey farmer considered himself fortunate in renting land at £40 an acre for early potato and tomato growing. A Lakeland farmer deplored an increase of rent which worked out at twopence an acre.

Guernsey, with a population of 36,000 on twenty-four square miles, depended on intensive cultivation carried to a fine art; while in Sutherlandshire, with a rural population of two to the square mile, a human being other than a shepherd or keeper is a *rara avis*.

Early spring flowers, banana-trees flowering in the open, eucalyptus, palms, and bamboos, everyone associates with the Scilly Isles, yet few realize that they also flourish as far north as Loch Hourn and Ullapool in north-west Scotland. You find these rare plants and trees in deep, narrow, sheltered valleys, warmed and kept frost-free by the Gulf Stream, with a climate as mild as the Riviera's, although within sight of snow-capped mountains.

In the fourth year of the Greater War good land can still be seen lying derelict in England, yet in the west of Ireland I saw soil being scratched out from between the rocks and carried in baskets to make a field, and a 'big farmer' might have two or three acres in three or four plots spread over the mountainside, and count himself lucky with a fourfold increase of potatoes, laboriously manured with seaweed and muck carried up the slopes on human shoulders. Here, too, I was assured by the local matchmaker that he could find me a good wife for £80, in fact I could have my pick. Mistaking my look of surprise he assured me that it would have all to be cash down, and the father would give his best cow as a wedding present. Remembering the old saying 'Loveless as an Irish marriage', I inquired if they ever married for love and not by arrangement. He snorted in disgust, 'Yes, in America, where the divorce rates are two in seven'.

Yet for all the contrasts, from Land's End to John o' Groats, from Lowestoft Ness to Valencia Island, in spite of all the differences in wealth, religion, outlook, and political opinion, the great freemasonry of a common calling means a welcome everywhere. In the house of a great landowner with a quarter of a million acres, and in a remote shepherd's cottage, I have spent many a happy hour discussing long and earnestly, in perfect harmony, the problems of the world's greatest industry.

On the completion of my tours in this country I turned my eyes to the Continent. Far too many people cross the English Channel before they know their own country. I have heard them making invidious comparisons between English and French farming. Had they studied the crops on similar geological formations they would see that the best type of British farmer has little to learn. A great statesman once said, 'What do they know of England, who only England know?' Very true! But what do they know of England who only the Continent know? Ask the man who comments unfavourably on English farming when travelling between Paris and London, if he has ever seen the Fens or Lothians. Invariably he has not.

There are, of course, many other things of interest in Europe besides farming and I have thoroughly enjoyed my visits there; and it does a farmer good to know that the world extends a little beyond his own boundary hedges and the nearest market town.

Every holiday brings its little adventures. I have had my fun in many odd places, while my brother had the good fortune to meet his young and charming wife on the Farmers' Tour to Canada and America in 1939.

Of course we never let our holidays interfere with our farming. We planned them well ahead, got the work well forward, and then felt we had deserved the break. With partners also, either of whom could run the business, it was easier to get away. It is a great advantage in farming, even if you are working seventy or eighty hours a week, to know that you can get away from it if you want to, and we made the best of our opportunities.

Since the war we have had neither the time nor the inclination to take even a day off, but we still have our cherished memories. One partner can say, 'Now when I was in Budapest . . .' The other 'In Toronto or New York . . .' What a lot you can get out of eighty-five acres of poor, stony land with a little thought and energy. The art of living consists in concentrating on the things you want to do, leaving the rest alone. The total cost of all our holidays was less than many a man fritters away in beer or cigarettes over a few years, though actually paid for by photographs and articles sold to travel magazines, for a farmer must never miss an opportunity of earning a little money!

# CHAPTER ELEVEN

## THE FARM BUILDINGS

Nearly every farm in the country has inadequate farm buildings for the area of land they serve, and it is quite impossible to stock and crop the land as it should be stocked and cropped without sufficient suitable buildings.

This farm was no exception; we had to adapt the buildings to our purpose, and as capital was so limited, do the building ourselves. This, I think, would have been the solution for many other farmers in a similar position. As an example in the difference of cost: when we first came here the house needed repointing, i.e. the old mortar between the bricks replaced by cement. We had a quotation from three firms, the lowest being £75; allowing 1s. an hour for labour (farm workers then received 7d.) the total cost including materials, when we did the job ourselves, was £25, and the work is as sound to-day as when it was done.

In planning we had five guiding principles. The buildings had to be adequate for the purpose, cheap to construct, convenient for use, lasting, and pleasing to the eye.

If others plan to build on similar lines in the future I should remind them that farm buildings now have to be approved by the local authorities before work can commence, and if a cowshed is involved it will be necessary to know the requirements of the County Agricultural Authority, as almost every area has a different standard. It is possible for there to be two cowsheds within sight of one another, identical in construction, but in different counties—one being passed for the production of the highest-grade milk, and the other rejected; which indicates that there should be a national standard. Personally I am of the opinion that the man in charge is of far greater importance for the production of clean milk than a five- or five-and-a-half-foot standing, but the County Organizers, never having milked a cow before 5 o'clock in the morning, know better, and it may be that Oxfordshire and Berkshire cows muck in a different place. In view of these variations we do not show any measurements on the sketch map of the buildings, as we do not wish to get anyone into trouble by using our dimensions, or cause our local authority any loss of sleep, to think hat in their county, of all counties, there should be a cowshed wit h the wrong-sized gutter. Though in actual fact our standings were designed for Jerseys, and after twenty years we would not vary them by an inch.

From the general description of the farm earlier in the book, and from the plan on page 141, it will be seen that the building originally consisted of a large Cotswold barn, with stable attached, and a long open shed, which on the west side served to house cattle lying in the large open yard, and on the east at one end making a cartshed. The barn and stable were stone-roofed, while the open shed was thatched. The yard seemed bottomless when we first cleaned it out, and a horse would sink in to his knees unless there was plenty of straw.

One of the first alterations was to put a wooden floor in one end of the barn (as we did not intend to use it for stacking corn). This gave us the same floor space for tipping grain, but also room to house the incubators, as the walls were very thick and with the floor above gave us an even temperature. In later years the other half of the barn has been fitted with corn bins for the storage of feeding stuffs, which in the early days had to be shot on the floor.

Then we built an open shed on the south side of the barn for cattle to lie in, as we intended to use the original open shed as a cowshed; this was done by fitting roof lighting and walling up the front, the thatched roof being replaced by corrugated iron—painted, as all iron and asbestos is on the farm, a dark green. There was sufficient width for a feeding passage, mangers, and the usual tubular standings for the cows, which were fitted in due course. As an example of difference in cost, a local builder quoted £30 to put on the corrugated roof, we to provide all materials—it took my brother and me exactly a week, besides doing our usual stock work.

As the herd increased in numbers, and it will be remembered that they had to pay for the buildings out of profits, two covered yards were added. Meanwhile we had dumped many loads of stone in the bottom of the large yard, with smaller material on top, and concreted over the whole surface, making a yard which could be cleaned up thoroughly. Then a bull box and calf pens were built into the old stable, for with the reduction of horses, they could be housed in loose-boxes when necessary. A new cartshed was built, the old one being converted into an isolation box, for it had no direct communication with the main buildings, and was later deemed sufficient under the Attested Herd regulations.

The silo is placed where it is easy to fill, either by hand or cutter blower, and close to the stock for feeding. Running water has been arranged for troughs in two of the yards, while there are taps in the cowshed, calf pens, barn, and isolation box, to save labour in carrying water, while the whole buildings area, including the road round it, can be swilled down.

The dutch barns are just across the road, convenient for hay and straw in the winter. One of them has been walled in on one side and one end, so that it can be used for a cartshed, or covered yard, when not required for corn, hay, or straw. We usually lamb the ewes there, for space is available in the spring.

The total cost of the alterations and additions, without labour, was £300. The dutch barns cost another £130, using asbestos sheeting on these adding to the cost somewhat. Levelling the site and building the barns involved five hundred man hours, more than half of which were spent on preparing the site.

The pig-house, mentioned in the chapter on pigs, cost just over £250, and took every spare moment for over a year to complete it. To level the site we had to move over a hundred tons of soil by wheelbarrow, but it was well worth the labour involved, for it enabled us to have a ramp at the end from which pigs and manure can be loaded without lifting, and meal taken in from a lorry as easily, while underneath is the liquid manure tank, from which thousands of gallons have been taken for use on the land. The system of lighting the house is unusual, being roof lighting along the whole length of the feeding passage; this makes the pens light and the dunging passage dark, which ensures the pigs keeping their beds clean. One visitor was shown round and everything explained, without his saying a word, but finally he exclaimed, 'Humph! Pigs in the drawing-room, but gosh they do look well!'

On the top of the pig-house is a water tank, which supplies the whole set of buildings, a 3/4-h.p. engine being used to pump the water direct from the spring, at the rate of 2,000 gallons per hour.

The brooder house, for rearing chickens up to eight weeks old, is constructed with asbestos and lined in the same way as the pig-house. Roof lighting is also similar, but in addition the whole of the glass front can be opened back to allow all the sun's rays to reach the chickens at all seasons of the year. With air-extracting ventilators in the roof and insulated walls, roof, and floor, it is possible to maintain the house at an even temperature, a most important feature if chickens are to be reared successfully. The materials used in the construction of this building cost £110.

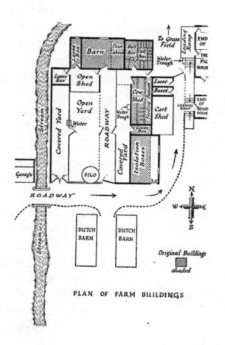

PLAN OF FARM BUILDINGS

At a total cost of £790 most people agree we have a good set of buildings, but by doing the work ourselves, and often having to leave a job unfinished for months, we were able to achieve our purpose. The expenditure was spread over a number of years, for it will be remembered we planned to live on the income from stock, leaving income from grain for debt repayment and improvements. Each class of stock had to pay out of paper profits for their own buildings. If we had used the money for general expenses, then the corn cheque when it came could be used for the improvement that particular section of stock was entitled to, according to the books.

The building of the road, across the swampy piece of common land and into the farm, was almost the biggest task we have ever undertaken. A distance of 660 yards involved the quarrying, carting, spreading, and breaking of over a thousand tons of stone. It took us ten years to complete, when we had a tarmac surface put upon it, and now it can carry the heaviest type of lorry, threshing engine, and recently, to our sorrow, heavy army tanks. What price could be put upon a construction of this description we do not know. A previous owner once had a quotation for £1,200, rather more than the whole farm was worth at the time, so it was never done. In our case we had only to give our time and labour, apart from tar and sand. To us, who have hauled out bogged carts from the swamp, had horses go in up to their bodies, and on one occasion spent a week getting a threshing engine from the council road to the rickyard, it has seemed well worth while. The trouble we took, draining away springs, deepening ditches, building three bridges, seems to have been well justified when a lorry and trailer can come in, turn round without shunting, and go out with twenty tons of potatoes or wheat.

In August 1939 my brother decided to get married in the following spring, and we therefore needed another house. It had long been his desire to build a house of his own design, so here was the opportunity.

Now we have always considered that no more money should be spent upon a property than it should be worth in the open market. You sometimes see an advertisement in the agricultural papers, 'Small Farm, £7,000 freehold. £20,000 spent on the buildings fifteen years ago.' Now Oathill, is not, and never has been, a rich man's hobby, but an economic proposition from the day we took over, so we decided £1,000 should be the maximum cost, and with any other improvements we have carried out, the whole property would be well worth our total expenditure over the years.

This was perhaps a little hard on my brother who wanted every comfort and convenience, such as central heating, large windows, cloakroom with heated cupboards for wet coats, hot and cold water everywhere, and other refinements. However, we have had a lot of practice in 'making the garment fit the cloth', and I had every confidence he would do it.

Had the times been normal he would probably have built it brick by brick, but we were very busy changing over our farming system from peace to war, as described in another chapter; we also had several other irons in the fire, so for the first time we employed a builder.

But first the rough plans were made and taken to a local architect, recommended by the Rural District Council, for without their approval nothing can be built, After this my brother consulted his fiancée, whose requirements were simple—it should be warm in winter and have plenty of cupboard space. Her life in a large Suffolk house (believed to have been owned by John Winthrop, who sailed with the Pilgrim Fathers to America) which was impossible to keep warm, and had few cupboards, undoubtedly prompted her ideas as to an ideal house. He also showed the plans to many of his friends, asking for their criticisms, and received several excellent suggestions.

By then war had been declared, but we had purchased all the materials, and had them delivered on to the site or into the farm buildings. The only part not received was the staircase, which a Midland firm had quoted for, claiming they could make up when required as they had the wood in stock.

We then approached the builders for quotations, materials in hand being contra-accounted against their quotations at cost. In other words they would be paid simply for building the house. Quotations varied from £1,420 down to £950. After inspecting two houses built by the man giving the lowest quotation we accepted the tender, specifying that the house must be completed by the following March.

The foundations were laid in October, but wet weather caused delay and by Christmas the building was only up to the ground-floor window-sills. Then the hardest winter we ever knew set in and work was stopped for eight weeks. The ground was so hard that we could not even dig the trench 300 yards long which was needed to take the water main. When the frost gave the brickwork had been damaged to such an extent that the 'house' had to be taken down again to the foundations. This was a great disappointment to my brother, as he always likes to keep to the schedule which he has in mind. However, a better house resulted from the delay, for the water system was replanned to guard against so severe a winter in the future, and no trouble was experienced in the following winter when we had another severe spell.

The position in early March, when the house should have been nearing completion, was the same as in October—i.e. only the foundations were laid. Another worry was that the small spring which was supplying the water for concrete mixing would run dry at any time, therefore we had to push on to getting the water main across the field. The trench was dug perfectly level at the bottom, through a hillside, so that there should be no air-locks in the waterpipe. This involved digging down to nine feet in parts of the field, some of the way through solid rock with a cold chisel and hammer, splitting off an inch or two at the time. How slow this seemed, with the other water supply failing. Finally we got through on a Saturday evening, the spring drying up on the Sunday.

The builder promised to put on several bricklayers, but never had more than one working; both the builder and his son were highly skilled carpenters but they were content to mix concrete and carry bricks, which any labourer would have been content to do at a shilling an hour, as they could not get on with their own work until the brickwork was more advanced. From our experience of planning work we knew they were wasting their time and money by not getting more bricklayers. Another annoyance was that they would not work a minute overtime, although they had not earned a penny for eight weeks.

Progress was slow, but as my brother was 'clerk of the works' nothing was skimped, and he had everything done thoroughly.

In late April we were ready for the staircase and wrote to the firm who had quoted for this, asking for delivery. They replied that they had used the timber for another job, and it would be necessary to get a permit to purchase.

Our application for this was rejected, but our long experience in dealing with Government departments was equal to the occasion. Three identical forms were filled in and dispatched, two of which were again rejected, but the third brought the permit by return.

The builder had been drawing money from time to time, which meant that to all intents and purposes he was working for us, leaving a little money in hand. But one Friday night he came for money to pay the plasterers (this trade being highly skilled was being done by subcontract), who wanted half the quoted price for the job. My brother asked if the work was more than half completed. The builder said he was sure it was. But no plasterers came on the Saturday or Monday. So the builder was instructed to see them, and was told they would not be coming again as they had got a job at more money in Oxford. We 'phoned the Labour Exchange and found they had plasterers on their books. So the builder was told to warn his plasterers that unless they were at work again by the end of the week we would have it completed by direct labour and sue them for the difference in cost. They promised to return on Friday, and did so, but this was another week lost.

*THE AUTHOR AND HIS BROTHER*

In late June the builder came to the conclusion that he could not make a profit on the job, due, we considered, to his inability to organize his labour properly in the early stages; for we now believed we could have made a profit on the figure by direct labour. Finally he told us one Friday that he would not be coming again, as he had joined the R.A.F. as a carpenter and had to go on the Monday. Fortunately my brother's varied experience of this class of work on the farm buildings was sufficient for him to complete the house unaided and he moved into the house in July 1940.

After three years' occupation, both my brother and his wife say they would not alter anything in the design or construction, so it may be of sufficient interest to give details here which may interest those who are thinking of building in a rural district after the war.

Being in the Cotswold area the house had to be in keeping with the limestone-built houses of the district, but as cavity walls were desired for dryness, stone-coloured brick was used, which from a short distance gives the appearance of stone. The roof is of hand-made sand-faced grey pantiles, which give the house an attractive appearance quite in keeping with the familiar Cotswold style.

All the timber used was Columbian pine, which after staining and polishing comes up rather like dark oak. All the floors are secret-nailed, so that no nail heads appear on the surface. Window frames are metal, with special hinges which permit the outside to be cleaned without leaning out of the window. The water is supplied from the never-failing springs at the farm buildings and goes by gravitation to the house. Water from this limestone soil is hard; so hard that a water softener is impracticable, and to overcome this a 300-gallon tank, with an overflow to the drains, has been built in over the coal store to collect all the rain from the roof. This closed tank, with an inspection plate for cleaning, appears sufficient for the purpose, for they have never been short of soft water. A pipe brings this water to the copper, which can be filled by a turn of the tap, which is sufficiently high to fill a jug or bucket as well if desired. There is also a tap at the bottom of the copper for emptying it; why most coppers have to be bailed out I cannot think, unless they were designed for women's use by men who have never had to do this unnecessary work.

Being two miles from a village and eight from a town there is no electricity, gas, or main drainage. This house has therefore been fitted with a large septic tank, which seems a most efficient method of draining.

The hall, cloak-room, scullery, and larder have buff-tiled floors; window sills for these rooms are also made from the same material. In the larder there is a white-tiled shelf, two feet wide and four feet long, in addition to seven eleven-inch shelves running along the two longest walls.

The living-room has a radiator in the bay window, under the window-seat, and even in very cold weather it is very pleasant to have meals at a table in the bay window, which has a good view over the garden and countryside; in fact if the sun is shining it seems like spring on a winter's day, as there is a gentle heat from the hidden radiator.

A radiator is also fitted in the hall, in a narrow recess under the stairs so that there is no risk of hitting it with a tea trolley or of a child running into it. Near the kitchen door there is a large cupboard with shelves and space for brooms, etc. The cupboard under the stairs is entered from the kitchen.

The cloak-room is fitted with a steel cupboard—hot-water pipe running under it—with holes in the bottom so that hot air rising through it will dry coats overnight. This is a special blessing for an outdoor worker in very wet weather. A compartment at one end has shelves for boots and shoes, the bottom shelf being used to dry them when necessary. A steel seat between the cupboard and the washbasin is so designed that Wellington rubber boots can be stored, for it does not do to keep them in a heated compartment. It will be appreciated from the plan that this room can be entered from the scullery or. the hall. The objection that if the latter door were left open it would give a visitor at the front door a view of the wash-basin and so on was overcome by fitting a door-closer.

The kitchen has white tiles behind the oil cooker and above this there is a ventilator, taken into the chimney stack to carry off the heat and smell of cooking. Provision was also made so that a range could be fitted at a later date if desired.

In the scullery there is a sink eight inches deep and the bottom is set thirty inches off the floor—the plumber thought it was quite mad to fit a sink so high—for most sinks are set too low, and why should a woman have to bend her back when working there? There is a draining board each side of the sink and below there are cupboards with shelves. The position of the copper and boiler will be seen on the plan (the recess in which they stand had been tiled, and anthracite coal can be shovelled from the coal store to the boiler quite easily). There is a clothes-dryer, on pulleys in the ceiling, running full length of the room and in wet weather the washing can be dried without trouble as there is a certain amount of heat rising from the boiler. This has been especially useful since the baby arrived. Between the draining board and back door there is sufficient space for the 'mangle-cum-table' to stand.

On the second floor, in number one bedroom there is a large cupboard— between the door and the bay window—with the usual wardrobe fittings, and above is a small cupboard for hats, etc. A hot pipe passes up one side, so that clothes are always aired. There is another small cupboard at the side of the fireplace for shoes, with four shelves in it. The radiator is in the bay. My brother could not find a fireplace to please him, so he designed one he thought his wife would like, and was, I think, much relieved when she did express her approval.

Number two bedroom has a small radiator under the window and again the cupboard has a hot pipe passing through it.

In number three bedroom there is a large radiator between the two east windows as this is likely to be the coldest room, for the east winds can be very cold on our side of the Cotswolds! Again there is a large cupboard with a small one above. There is also a fitted basin in this and No. 2 bedroom, which my brother added in completing the house, at a cost of less than £10, for it occurred to him that with the fitted basin only in the bathroom there might be some delay when visitors were washing. It may have also occurred to the reader that we both have a horror of wasting time!

The bathroom has the usual fittings, including a hot towel rail and a large airing cupboard, which does not have the hot-water tank in it, for tanks often take up too much room in airing cupboards, but pipes passing backwards and forwards across the back. The hot tank is in the scullery above the boiler and copper, as the greatest efficiency is obtained by having this as close as possible to the boiler.

Above the landing there is a trapdoor into the roof space, the ceiling rafters of the bedrooms being boarded over; this makes a good store for apples, etc. The roof is felted, under the tiles—there are also hot pipes crossing, which prevents freezing. These pipes are lagged, but even so there is a certain amount of heat from them.

The reader may wonder why there is no garage, but as they will not want one for the duration of the war this was not included, though it will be seen on the plan that space has been provided for it. If by the time the war is over and the world settles down, the family plane is the common mode of transport, then they will only have to build it with a door in the back through which to wheel the plane from the field, instead of driving the car up the driveway. It will be noticed that we still plan for the future!

I wonder what faults other people may be able to see in the plans of the house, or from my failure to describe the fittings? But I do suggest that it is a very comfortable and convenient place to live in, one which should be within the reach of most people who require a three-bedroom house after the war, on the small new farms which I sincerely hope will be a feature of the English countryside.

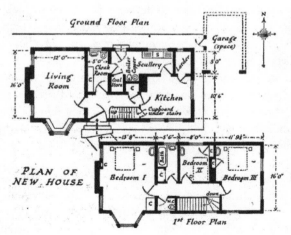

# CHAPTER TWELVE

## THE FOURTH RUNG OF THE LADDER

In describing how I set myself to become a farmer in five years, and achieved my ambition in four, of how we planned to establish ourselves as tenant farmers in seven years and owner-occupiers in a similar period, both of which were accomplished in ten, it would have been very nice now to be able to say that we intended to become agricultural landowners with tenants of our own, as part of the great plan. Unfortunately that is not true; fourteen years is a long time to plan ahead at the age of twenty. We did not realize then how rapidly the years slide away, and that we would have to do something with the rest of our lives.

No. We climbed into this much-despised class fifteen years after starting to farm, and nineteen years after I left school. So within twenty years I travelled all the way from labourer (when I was cowman in Essex) to landlord. My brother, being two years younger and a farmer before he was eighteen, has done that much better. Never let any young man say there are no opportunities in farming. They are there all right, but mostly disguised as hard work. Opportunity, we are told, only knocks but once. Personally, on looking back, I realize she spends most of her time hammering on someone's door, but they do not bother to get up and let her in, or complain the door is too heavy, jammed, or pinches their fingers!

Of course, we have continued to farm our own land, but the ownership of farms which are let to good and careful tenants can afford an equal satisfaction. With a good income from our own farming, rents can be reinvested in the land from which they are drawn, and we can look on property as something which appreciates, in value over the years when it is not bled white by the taking of rent for dissipation elsewhere.

Our method might find other adherents, and remove the reproach against the landlord system, if each landlord farmed a small area really well, preferably his poorest farm, and lived on the income from it, reinvesting his rents for the improvement of his estate. No tenant can whine, "Farming is doing badly' if the landlord himself can show that the land he retains in hand is a real business proposition. The same applies to the Church; the moan of the poor clergy would no longer be ventilated in the columns of the daily press if they kept their glebe farms (usually about the same area as Oathill) under their own direction, and made a few hundred a year from their farming. Or if the Ecclesiastical Commissioners worked a few of their farms on the vast estates from which they draw rents but return the least possible amount in repairs.

In many respects this aspect of our career has been the most difficult, for in taking over tenants from a bad landlord it takes a long time to win their confidence. They always suspect that a fellow farmer has bought with the intention of turning them out and farming it himself; or they have been misled by promises that repairs will be put in hand, but which are never done; and what they hate most of all is a landlord who wants the land farmed as it should be farmed.

In each case we have had to re-educate our tenants, or turn them out and replace them by men after our own heart, before we can do anything with our land. It is also very remarkable that the War Agricultural Committees will seldom help a landlord to have the land farmed as it should be. Being mostly large-scale tenant farmers, they are bitterly opposed to the landlords and suspect that they are trying to get the Committee to do their job. In one case we sought their help, where a tenant we had taken over was going from bad to worse, but to read through the file of letters on the matter shows a sorry record in declining to co-operate in any way, and we finally appealed direct to the Ministry for permission to get rid of our unsatisfactory tenant, which was granted. The farmer we put in has grown more wheat and potatoes, produced more milk, the three most important commodities sought by the Committees, in one year on this badly neglected farm than the previous occupier did in three. Another of the mysteries of tenant farming is the man who wants to hang on to land, from which he is not threshing sufficient corn to pay the rent, where he is always behind with his work, and always having trouble with his labour. It must simply be that he wants to boast of the area he is farming, though it could give him little pleasure to show his friends round it.

Our general rule has been that now that half the rent goes in income tax, one-quarter is spent on repairs and improvements, and another quarter is kept for reserve. Therefore letting a farm at five per cent interest on the purchase price leaves us one and a quarter per cent as interest for the reserve fund, which we hope to reinvest in due course. As long as it was possible we carried out all repairs by direct labour at half the cost of builders' quotations, so that the tenant had twice the value in repairs or improvements, for the quarter rent he paid to us. Gates, etc., were made in the workshop at Oathill, while a number of 'estate' fencing jobs were done by our staff.

We realize that good landlords and good tenants are as rare in this country as good labourers. But when two or three can get together, then things improve. We as landlords are determined that every help shall be given to any tenant who is doing his best to maintain maximum production and not robbing the land. This is where our pupil system comes in. Any boy who has really proved his worth and learned our methods can be trusted with land and capital, and nowhere could we find sounder investments. When the time comes for us to retire, how nice it would be to have seven or eight well-managed, well-let farms, where the methods we have proved are being faithfully applied. This would be far, far better than trying to manage a large acreage in our declining years, which is the mistake too many capable farmers make, when hanging on to the reins too long.

# CHAPTER THIRTEEN

## WARTIME FARMING

The crisis of 1938 found us quite unprepared for war. The writer had always believed that Europe would never be led into war while the leaders of the nations could still remember 1914 to 1918. Surely even Hitler, who was reported to have been temporarily blinded by poison gas in the last war, would never permit the youth of Germany to suffer as he had? In our travels, the people of all countries seemed kindly, decent folk; and war was just a nasty memory of our youth, and the sensation of the moment in the popular press.

How wrong I was, but that crisis did give us a chance to prepare. In peacetime, although our farm was still two-thirds arable, all the corn we grew was sold for seed, and replaced for feeding by cheaper local grain. Purchased feeding-stuffs were delivered in weekly, and we usually had a fortnight's stock in hand. Compared with the Agricultural Returns for the whole county we were carrying three times the cattle, four times the sheep, ten times the pigs, and twenty-five times the poultry on our acreage. True our yield of grain and the stock-carrying capacity of the grass was much higher than the average for the district, but how were we to face a war?

First of all we transferred our custom from a port miller to a local firm. Then we steadily built up our stocks of feeding-stuffs, dry, hard grain especially, so that by the outbreak of war in 1939 we had a year's stock of food in hand. Hearing that Germany had doubled its orders at Aberdeen for cod-liver oil in the spring, we had also ordered sufficient to last a long time. Bought at three shillings a gallon, the price rose to ten shillings at the beginning of the war, and the oil was at that adulterated with whale oil. This stock lasted until early in 1943 without losing its vitamin content, proving a sound insurance against disease and a first-class investment.

We installed a mill for grinding, a silo for the preservation of green food, also a grain-sprouting cabinet, so that starchy food like barley could be converted into a valuable protein with double the value of early spring grass, at any time of the year, with water to which nitrogenous salts had been added. A pressure steamer for cooking pig potatoes and other waste was also added to our equipment.

We spent our spare time studying very closely all the available literature we could find on the substitute foods which had been used extensively on the Continent during the last war. Acorns, beech nuts, horse-chestnuts, weed seeds, hazel catkins, pine and fir needles, and other apparently unpromising material had been successfully used, chiefly by breaking down the cellulose content, and getting rid of resin, turpentine, and other undesirable constituents. We were amused when the straw pulp process was introduced later as something new and original.

Then came the war. Through all the difficult times before livestock rationing was introduced not one of our birds and animals missed its proper food; not one of our friends and neighbours can say we failed to help him during all that time if he was short, as we always had something we could spare. We must also acknowledge the kindness and help we received from our 'big' farming neighbours. At that time a merchant was allowed to sell back to the farmers one-third of the grain he bought from them, and a farmer who did not require it himself could oblige a friend if he was so disposed. Only one man refused to help us in this way, and told us we should reduce our stock. The others without exception went out of their way to help. One man we approached said, 'You know where my granary is, help yourself, and either return it when convenient, or send me a cheque.' Another went all the way to Banbury, fifteen miles, to bespeak some grain for us, which he had already sold. With what pleasure all the resources of our workshop and technical knowledge were put at the disposal of the first man, when he broke a machine and had been told it was not repairable since necessary spares were in America, and his crops were spoiling for lack of it. Back in the field in six hours, he asked how much he owed us. We gave him his own reply. 'You know where our "granary" is, help yourself.'

For the man who went to Banbury, we wrote up a complete set of books when he received an unjust demand for income tax. If money saved is money gained, he earned about £100 a mile when he made that journey to Banbury out of the kindness of his heart. God bless him! When we asked for the grain, he said, 'Yes. It would be a thousand pities if you had to reduce your stock, after you have tried so hard to build it up.'

There are some who helped us then, whom we have never had the opportunity to repay, but they will find us willing to help at any time they require it and all the resources of Oathill will be at their disposal, with money, labour, stock, or grain should the occasion arise.

Why should we maintain our stock? Our reading had shown us that Germany lost the last war not through lack of grain or potatoes, arms or men, but of livestock products—fats and proteins, without which the health and morale of the people is steadily sapped until they begin to think, 'a horrible end is better than a horror without end'. They lost the war for lack of bacon, milk, and eggs. I often noticed when I was learning farming, how much better I could work on the liberal diet in the farmhouse against the bread and onions of the poorer labourers. Also our records showed that the more stock we kept, the more grain we grew, through the utilization of their waste products. The manurial residues from our stock in pre-war days were equivalent to six tons of kainit, eight tons of superphosphate, and sixteen tons of sulphate of ammonia, and in a far more valuable form. In no way could we accept the Government's policy of despair, 'Scrap your stock and grow corn'. And the policy unfortunately seemed to be shared by our local War Agricultural Committee. Break pastures by all means, but you can still keep as much stock on the arable, or even a little more.

Once the rationing scheme was established and We could obtain our fair share, the feeding problem became much easier, as we knew exactly how much substitute food was required to maintain the stock.

Taking a forage crop for silage from the land before it went in roots enabled us to do without purchased cake for the cattle and sheep; this alone represented a saving of ten tons of concentrates. It has also made possible the production of nearly £5,000 worth of valuable dairy stock, which would have been lost to the country had we not adapted our system to the times.

The silage campaign has not received the support it deserves. We have learned to value silage so much that we no longer look upon it as a mere wartime expedient, but as something well worth incorporating in our general farming practice for the future. In comparison the old methods of feeding now seem wasteful and extravagent when such an excellent substitute for expensive cake can be produced economically and easily on our own land. It is also safe and weatherproof, fitting in well with the ordinary routine work of the farm. The slogan of the Ministry, 'Make silage, make sure', has been proved here this year, when kale has been a disappointing crop, but we have the silo full of good stuff, without which we could not face the winter with equanimity, for silage will fill the bellies while supplying the protein the cattle require.

There is quite a lot of work involved in silage making; our records show that it takes twenty man hours to harvest an acre of vetch silage, against four man hours of vetch hay. But surely those sixteen man hours are justified, for it doubles the units of food obtained from each acre.

Analysis of samples seems as misleading as analysis of the soil. We have had reports varying from 14 to 23 per cent protein, without being able to detect the slightest difference in its feeding value as judged by the health, growth, and condition of the stock.

We take extreme care in the making of our silage, far more than the experts recommend, but we think it has been well justified. The concrete silo cost £75 and saved a £100 cake bill the first winter, had the cake been available, which of course it was not. When I think of the good stock we have reared on silage, I sometimes wonder if we should not raise our hats to the silo, very much in the same way that naval officers salute the quarterdeck!

Steaming potatoes, up to fifteen hundredweight per day, makes a lot of extra work, but it has been well worth while for it has enabled us to maintain the pigs, which are now in demand again since the edict has gone forth that pigs have been reduced far enough, and they are now required to consume the waste.

Mangels have also proved a very valuable crop, and we have exerted all our skill and knowledge to make the best of them. We have grown up to sixty tons to the acre, stored them with extreme care, so that they tide us over that period in the summer when pig potatoes are not available, or of such poor quality that they are not worth the cleaning-off of sprouts, which are harmful to stock, before cooking.

Catch-cropping has always appealed to us, and our practice of mixing trefoil with the seed when drilling oats in our early days to provide stubble grazing for geese, still stands us in good stead, for it is now grazed by sheep or cattle.

The extent to which land can be fully occupied is illustrated by the field I can see as I write. Winter oats were cut in July 1939, and the stubble was mucked and ploughed between the shocks of grain. It was drilled in rape on the 13th of August, and was penned off with sheep in December, and ploughed immediately. Then came a sharp frost for eight weeks. In early March vetches and oats were planted on a good tilth made by the frost, and the crop was cut for green food and silage in July. Kale was planted on the 21st of July and by mid-October we were cutting a cartload per day from ten yards by ten yards, or forty-eight loads to the acre, the sheep clearing up the second growth. Then oats were drilled in mid-February, with clover and ryegrass in the bottom.

After a good crop of oats, the stubble was grazed, and then shut up till spring. Dressed with liquid manure, grazed off with sheep, then a hundredweight of sulphate of ammonia was sufficient to give us thirty-three hundredweight of good hay to the acre in June, the latter month being grazed by cattle. Then the field was mucked again, ploughed, and put in wheat, trefoil being broadcast with the artificial manure in the spring, for grazing this autumn. So that in four years we have grown seven crops, and also provided grazing for sheep and cattle between on land upon which the poultry also run for about half their time, for they obtain great benefit from stubble, ploughing, and green crops generally.

Once our system was modified to war conditions it became once more a matter of routine, and although we have worked far harder than ever, never less than eighty hours a week, we have very few worries. 'Black-out' is difficult, for we belong to the 'lantern farmers' whom so many writers on agriculture fail to understand. Before the war our neighbours looking across the hills would see a blaze of light, Tilley 300-candle power pressure lanterns, and say, 'There's Oathill still at it'. Under war conditions these have to be shaded, and this hinders us considerably. It is a remarkable thing that so many farmers will work late in summer, but stop at dusk in winter simply for want of efficient lighting. There is hardly a job we have not done at night, and there is no hour of the day or night which I have not heard Kiddington church clock strike out when the wind is in the south.

The answer to your question, dear reader, how does a farmer find time to write a book when working eighty hours a week, is simple; this has been written between the 9th of May and the 4th of September, on Sunday afternoons, between one o'clock and half-past four, when most farmers take a well-deserved nap. I had previously prepared an outline of the book and submitted it to the publishers, and on the 7th of May was asked to get it done by the autumn; and once I am set a task, I am completely happy until it is finished, when I look round for something else to do, though in an undertaking of this description it would have been nice to have the leisure of a professional author or civil servant. When I read of, say, Arnold Bennett, a master of English, writing 500 words a day, I cannot help thinking how nice it would be to spend an hour pruning and polishing each paragraph, for this book is written at the rate of over a thousand words an hour and even then my mind runs far ahead of my clumsy fingers; for I cannot write to a synopsis, and use the simple narrative form, as I have told it in far greater detail to my pupils as we work together in the fields.

However, to return to wartime farming. Apart from the 'black-out' petrol is our chief difficulty. The powers that be do not realize that intensive farming requires far more petrol than general extensive farming. What a wicked waste it is that one often has to keep a tractor running on paraffin when muck carting (we have six hundred loads a year) when a drop of petrol would start the tractor again. For T.V.O. takes up quite as much shipping space as petrol. How we hate to see it being wasted by the Services on unnecessary travel when we could use even an extra pint to the best advantage!

Double summertime and ordinary 'summertime' in winter make no difference to us. Providing the cows come in to milk just before sunrise on the 21st of June, they do not seem to mind a bit whether our clocks say 3.50, 4.50, or 5.50, for they will be milked again twelve hours later. Changing from summertime to wintertime means again an alteration in the clocks but none in the star time for feeding and milking. Childish devices for getting the townsman up early mean nothing to us or our cows.

The farmers' moan about form-filling leaves unmoved anyone who has a proper system of bookkeeping. The amount of work involved is nothing compared with keeping proper records and pedigrees. We ourselves have two systems of bookkeeping, a very simple one for the Agricultural Research Department of Reading University, and a more complicated method of our own built up over the years which enables us to detect any weakness which may develop or budget our expenses ahead if we wish to tie a lot of money up in some new venture. It also enables us to make the very best use of our land. Technical knowledge and practical experience are not enough; co-ordination, organization, and administration are also essential, and these are only brought together by accurate accountancy. In our early days I grudged the few guineas my brother spent on auditing; I now realize that his business training was the third leg of the stool on which our little edifice has been built.

No chapter on wartime farming would be complete without some reference to the War Agricultural Committees. We deplore that they should have been considered necessary; it is a very great reflection on British agriculture as a whole that each individual farmer was not prepared to make the best possible use of his land in the national interests, which are, of course, identical with his own. The greater the production the greater the profit; the more the farm is self-supporting the less you have to spend. What a mistake it is to excuse inefficiency on the grounds that 'farming is a way of life'. One should certainly farm because one likes it, but they will like it a great deal more who also make it pay well, and without profits to reinvest nobody can make the best use of the land.

Our relations with the local committee have been far happier as farmers than as landlords, though individuals can be very annoying in their manner of carrying out their duties, and in expressing opinions which are outside their province.

In the early days of the war a member of the committee called, and in looking round said that it was the small farms which were not fully productive. Now this is quite contrary to the writer's experience, in visiting over a hundred farms in my travels before the war, and my view is supported by official statistics. I instantly made the following offer, or challenge, which is still open.

'If any large-scale farmer on similar land could produce properly audited and analysed accounts to show over a longer period a higher output per acre, per person employed and capital involved, we would pay £100 to any charity. Also if it was contended that our high prewar output was due to cheap, imported feeding-stuffs, then if any large-scale farmer, in the Cotswold area of stone brash soil, could prove better figures than us on any recognized system of rotational farming, under wartime conditions, on the same basis, then another £100 would be paid to any war charity.'

Should any advocate of large-scale farming in this country like to earn a useful sum for his favourite charity, this is an excellent opportunity. Our output in 1942 was £90 per acre, £1,850 per person engaged, the gross annual return being four times our valuation of the live and dead stock. Twenty years of farming have enabled us to increase our output thirteen times and our capital a hundredfold.

But it is not the odd opinions held by members of the committee that concern us most. Freedom of speech was one of the Four Freedoms in which President Roosevelt crystallized the needs of the world—freedom of Speech, of Religion, from Want, and from Fear. I wish with all my heart he had included freedom from Bureaucracy, for in the committees this is found at its worst. In the last war a lot was said about farming from Whitehall. This was at least impartial, which is more than can be said of local control.

What is most annoying is to have an order served upon you compelling you to carry out something, with all the force of law behind it when this something is what you are only too willing and anxious to do. I give three examples.

In the early days of the war, after the farm had been inspected, we were told we were making the best use of the land, and no orders were made. Then the following year, at the time of the great survey, which actually took nine minutes and consisted of writing on to a map the crops I said were in the fields, I mentioned that we intended to plough up fifteen acres of ley, even though it was a first-class plant and a mass of white clover, being the finest Aberystwyth strains, as we could carry more stock with oats and forage crops. A few days later we got an order to carry out this work, which had in fact already been started. Now why should we be compelled to plough good ley, when in 1943 there is on the adjoining farms still a far greater area of tumbled-down grass from the 1890's?

Late in January 1942 the local member and the area officer called to inquire if we would grow two acres of potatoes. I told them that our root land was already planted with vetches and rye, to be followed by kale, but if potatoes were required we would sacrifice two acres reserved for oats, but I did hope they would not make an order. The seed was ordered, the land baulked, and in due course along came the order. We grew an excellent crop, which paid us well, but lost three tons of oats, for the rest of the field yielded ten quarters to the acre, which we could put to better use in feeding stock. But what annoyed me was that we should be 'compelled' to grow potatoes, when we were already taking two crops from our root land, and another farmer in the district with 850 acres, 130 of which was bare fallow, some of which was only mowed, was not apparently required to grow a single potato. How much wiser, and more in the country's interests, to have made an order for us to grow the crop on the other man's land, which we would have been quite happy to do, for we did in fact do the cultivations on eight acres for another neighbour as we have a rigid toolbar for our tractor.

Then in August 1942 we wanted to disc-harrow and re-seed a rough steep bank of four acres. We had tried to improve it by heavy stocking, but thought we could do better. So we sought permission to carry out the work, which is required by law. Two members and an official called, and after inspecting, inquired if I did not think it was a waste of Government money. I said only our own need be wasted. We only wanted permission, not the £3 an acre grant. But the order was made, and so we felt justified in claiming the grant, which we obviously did not need, for the grazing in the first year alone has been well worth the expenditure.

With this steep bank we tried rather an interesting experiment. The grass came well, and was grazed back hard in the winter, on the advice of an expert from the Government Grassland experimental station. Then it was shut up till spring. Now many people will be familiar with the unit system of estimating the value of grazing. One sheep for one day is a unit of grazing, other animals in proportion. So early in April, on the day that a neighbour turned his sheep into old tumbled-down pasture, I calculated his grazing units per acre, and for every one he had on the field I put in ten on our re-seeded pasture. In mid-May I increased it to twenty units to one on the adjoining field, for the grass was growing away from the stock. In June we took out the stock, left it a week, and baled five and a half hundredweight of good hay to the acre. How well the 'compulsion' of the committee was justified. But why are there still hundreds of acres of old worthless pastures still awaiting their orders?

However there is also another side to the committee's work, and in my opinion of far greater value than the issuing of bureaucratic orders, and that is educating the farmers by means of demonstration farms. We ourselves have learned quite a lot by taking advantage of this opportunity to visit some of the best-managed farms in the county.

This farm was included in the list, and it is rather interesting to note the number of people who have overcome the difficulty of visiting an isolated farm under wartime travel conditions. If the response to the committee's scheme was the same on the one hundred and fifty odd selected farms, it may have been well justified.

Twenty-seven farmers, three from outside the county, ten students, two foremen, two land girls, and three ordinary farm workers took advantage of the chance to see something of our farming, and this is apart from several townspeople and foreigners who heard of the scheme and wanted to see something of English farming.

The committee also did us the honour of describing our farm in their monthly Farm Notes, issued to all the farmers in the county, only three other farms have been written up in a similar manner. As a fair comment on our farming I think it is worth reproducing, and I acknowledge the source.

## 'SMALL-SCALE INTENSIVE FARMING AT ENSTONE

'The small party of farmers who attended the demonstration at Oathill Farm, Enstone, on the 6th of June, were very interested in the degree to which fertility has been built up and the intensity of production resulting from it, in spite of the naturally poor, stone brash soil.

'Some striking comparisons were made with the Agricultural Returns for Oxfordshire; showing that, although the percentage of arable to grass has always been higher than that for the whole county, the farm can carry three times the cattle, four times the breeding ewes, ten times the pigs, and twenty-five times the poultry for the acreage compared with the pre-war figures compiled by the Ministry of Agriculture for the county. The output of corn resulting from such heavy stocking, producing some six hundred loads of manure annually, was also impressive, including one hundred and eighty-one quarters of oats from twenty-six acres in 1941.

'The general balanced system of farming on the holding remains the same in wartime as in peace, except that purchased feeding-stuffs have been replaced by home-grown cereals and good-quality silage.

'Every inch of available land is utilized to the full and cattle are tethered in some instances to make full use of odd corners and hedge bottoms and also arable forage crops. All liquid manure from the sheds is conserved and used on

the land and it is typical of the system that even the ensilic juices draining from the silo are caught and re-used for making more silage.

'A five-course rotation is practised. Fallow-corn-corn-one year ley-corn. Every opportunity of catch-cropping is taken. The fallow section, taking a winter-sown forage crop for soiling or silage, or vetches, rye, etc., is followed by kale. Stubbles are also cropped with Westonwolths Ryegrass or Rape. Sufficient manure is made to muck half the arable each year, and every endeavour is made to get it direct from the covered yards to the land, it being the rule that muck carted during the day must be ploughed in at night and never left in small heaps. The rise in humus content of the soil is considerable and instead of fearing drought, as with most thin soils, the records show that the dry years give the highest yields.

'The aim regarding stock on the farm is to maintain nothing but good-quality pedigree animals. Flocks and herds have been built up from quite small beginnings but Messrs. Henderson have always kept as their goal the establishment of herds and flocks disease-free and acclimatized to the land.

'Several farmers commented favourably on the skill and knowledge of the staff, which consists of three pupils, none of whom has had more than eighteen months' experience of farm work, yet help to maintain a gross output of some £1,025 per person employed. The system of farm apprenticeship existing here might well commend itself to other farmers, a system whereby keen boys of good education receive a thorough training in every branch, are remunerated on a profit-sharing basis, put in charge of a section of the stock, and finally given every help to obtain a good situation or take a farm of their own.'

Thus does a trained journalist sum up in 500 words that which I have taken many thousands to describe in this book, and all from a brief speech and a walk round the farm.

It is on the question of the powers possessed by the committees that the majority of farmers feel most bitter. While no-one can doubt that the vast majority of the decisions are made honestly and fairly, it is a negation of British justice that there is no appeal. It is well perhaps, or symbolic of the times, that the authorities of Lincoln Cathedral have sent the finest and cleanest copy of Magna Carta to America. Even the tattered specimen at Oxford has been hidden away; it is, I think in my more bitter moments, for fear some farmer should read 'None shall we depossess, From none shall we take away . . .' Star Chamber methods can achieve little, and it is a great pity that a principle of British justice is withheld by the Minister.

The practice of forming decisions from a casual inspection of farms is unfair. I should like to see the Hampshire system introduced whereby the efficiency of a farm can be assessed on a percentage basis determined by a

simple bookkeeping system. This would also remove a source of great irritation
to many farmers in the classification of farm into 'A,' 'B', and 'C'. Many resent
being in class 'B' when all the members are class 'A' regardless of the standard
of farming they maintain.

If the committees are to be kept in being after the war this Hampshire
system alone could bring about a great improvement in British farming. The
average output for each district would be known. The average farmer would
be class 'B'. Those with an output of twenty-five per cent above could be class
'A', and those with an output of ten per cent below the known average could
be in class 'C'. Then if only class 'A' farmers were allowed to take farms which
become vacant the standard would steadily improve. In the case of young
farmers taking their first farm, a fair condition would be to furnish proof that
they had worked on 'A' farms for say three years. For there have been far too
many people taking farms in the past without sufficient practical knowledge
or experience. To encourage good farming, 'A' farmers could be entitled to say
three years' rent if they received notice to quit, 'B' farmers to only one, and 'C'
farmers to none at all. For it must be remembered that it is the man in charge
who is classified, because success depends on the farmer far more than the farm.
A capable man will so adapt his methods that he will thrive on a few barren
acres. Another lacking skill and knowledge will fail with every advantage
showered upon him. In my travels through every county in the British Isles,
even at the bottom of the slump, there were a percentage of farmers who were
consolidating their position or going steadily ahead. There have been in the
past far too many so-called 'farmers' who by their own ability would never
have been more than indifferent labourers, and with the gradual elimination of
those who fail to become 'A' farmers, a better chance would be given to those
more worthy people who should have the privilege of farming the land.

As farmers we have no 'territorial ambitions'. We have everything we want,
but we should like to see a little more *Lebensraum* for those who are capable of
farming the land, without 'the extension of projective corn subsidies extended
to cover all main commodities produced by the farmer, together with price
insurance guaranteeing cost of production and margin of profit', which is so
vociferously maintained by certain sections of the farming community.

Is it to be wondered at that we have never joined a union that contends
that 'world conditions are such that farming in this country cannot be self-
supporting, and the policy of protection and subsidies has not been in operation
long enough to bring about improvement in methods of weaker farmers'?

Much has been written about depression in agriculture and that it must
be avoided at all costs in the future, but few realize that it is the depressions
that give the weaker farmers, such as ourselves, the opportunity to get a start.

Slumps eliminate the lazy, inefficient, and sport-loving farmer, together with the man who wants to devote his time to politics and local government, giving a heaven-sent opportunity to the man from the poorer and harder districts of the north and west to take the better land in the south and east of these islands. Also it is the opportunity of those of us who were reared in the towns, but regard the land as our birthright, and which we are prepared to earn in spite of every difficulty which may be placed in our way. A definite migration can be traced in the tenancies of farms, showing that this does take place. With two or three exceptions every farm in this district has changed hands in the last twenty years, the old Oxfordshire families being replaced by those brought up in a harder tradition. Ability to farm the land should be the main qualification for holding it, and a man who can take a farm with limited capital in a depression and make a success of it should not be kept out, at the expense of the taxpayer, by a man whose sole qualification is that his grandfather farmed it before him, and who is unable to adapt his methods to the times.

However, that is for the future. For the present we are thankful that we have been able to maintain the stock, which we considered to be our war work. This year there is every reason to believe we shall achieve the fifteen per cent extra production called for by the Minister for 1943, and if all-out production is required in 1944 or 1945, we trust our stock and reserves of fertility will be equal to the demand. We are at least gratified that the Minister's policy for a four-year Plan from 1943 coincides with our own for 1939. 'Taking the world as a whole there will be a considerable lack of stock, oils, and fats. We consider the best contribution we shall be able to make to the relief of hunger and distress throughout the world is to go on producing the maximum amount of feeding-stuffs from our own soil as long as this appears to be necessary.' The statement was cheered as the most hopeful yet from the Government. In my more modest moments I often think the Government would save themselves a lot of worry if they gave me a tinkle on the telephone. Our policy for 1924, described in this book, would be equally applicable to British farming as a whole, for 1944 or any other year, for it has stood the test of time. If each farmer had enough stock to live on and pay his way using the corn sales for improvements, the fertility of the soil would be maintained by the heavy stocking, while the heavy stocking would provide the population with all the corn, meat, milk, and eggs they require. For stock, through its manure, gives corn, while giving the other food required as animal products. If corn prices drop, then livestock is once more the mainstay of farming, for the low corn prices are more than offset by cheap feeding-stuffs on 'the farm which is heavily stocked. In the summer of 1943 the Archbishop of York was referring to 'haunting doubts' as to the future of farming in this country. He need, have none if the farmers used this period of unparalleled prosperity to put their farms in order, and stock up with healthy, disease-free, and good-type stock. To depend on corn is to live in a fool's paradise, as many found after the last war.

# CHAPTER FOURTEEN

## ACCOUNTS

Public liability companies are required by law to publish annual accounts and the directors are usually pleased to let the general public know how well they have done in the preceding year; but private traders, and especially farmers, are very reluctant that anyone should know their profits. Is this because they fear their employees and customers Will think that their margin is too great? It is also possible that they feel rather ashamed of a small profit, or fear their friends will think they are boasting if they mention a good turnover.

We have sometimes shown our accounts to fellow farmers, and in nearly every case they say, 'You must never let these figures be published. It would give people, and townsmen in particular, the wrong impression of farming. If an eighty-five acre holding can show profits running into four figures, it will be assumed that large-scale farmers are making a fortune, could pay higher wages, and do not need subsidies or protection.'

We do not know the profits in large-scale farming, but we do believe that if sound mixed farming methods were used, and a really high output maintained, highly skilled and efficient labour should be able to command wages equal to those paid in industry.

Unfortunately there is neither high output nor efficient labour on the majority of farms, large or small, and the farmer is caught in a vicious circle. Exhausted land and indifferent labour mean poor crops. Poor crops mean little food for stock, or low return by direct sales. Little stock means little manure. Little manure means poor crops, and the circle is complete. Without a properly balanced system, which keeps labour fully and profitably employed at all seasons, lost time is a ruinous drain on resources, resulting in insufficient capitalization of the holding. Lack of labour, where the farmer cannot afford a full staff, involves the loss of crops, or their serious depreciation, in bad seasons, and so once more the cycle goes round.

Now we publish our accounts simply to show what can be achieved when the farmer has solved the problem of maintaining intensity of production, with fertile soil and good labour; also to confirm our faith in the small farm, for we believe the future of Britain lies in the small farm, and this has been proved over the years. Between 1871 and 1931, while the numbers of farmers and their relatives remained constant, the paid labourers decreased by almost half. In 1870 more than half of England was owned by less than 150 landowners; by 1930, more than half of England was owned by nearly a hundred thousand small or smallish farmers. The labourers will continue to depart unless the system is changed, and after the war quite a high percentage of those at present on the land will have to be handed over to Sir William Beveridge, while young men will not take up agriculture without better prospects than in the past.

The only solution that suggests itself to me is that the best labour should be given the opportunity of becoming farmers, on capital earned on profit-sharing farms, which can afford to pay good wages as the result of efficient work and high output. Real intensity of production is only possible on small farms, and many of the larger and therefore unproductive farms could be subdivided for the benefit of the workers who have proved their ability to take a farm. This has been done in Eire, and figures shown to me indicate that production per acre has increased by three times, making land profitable which hitherto had failed to pay its way. I do not believe in a free handout to any class, whether farmer or labourer, therefore I suggest that the capital with which to take the new farms should be earned first, for in earning one also learns.

Our output of £50 an acre in 1939, and £90 in 1942, as an example of the productivity of a well-managed small farm, compares favourably with the figures quoted in the press for some of the well-managed, large-scale farms.

I have seen the output of the Boreham Estates quoted at £100,000 from 3,000 acres. Probably this is the finest example of co-operative profit-sharing farming in Britain, though it is unfortunate that the original outlay has not been disclosed; this would prove whether it has been a business proposition from the start. It will be noted that the estate has an output of £33 per acre.

Turning to large-scale private enterprise, figures quoted by 'Peterborough' in the *Daily Telegraph* on the 5th and 15th of June, 1943, show very much the same return, but falling output as the acreage increases. Other figures are:

The Hiam Estates, in the Fens, 7,000 acres, £220,600.
Mr. Rex Paterson, Hampshire, 11,000 acres, £200,000.
Messrs. Parker, Norfolk, Lincolnshire and Leicestershire,
    30,000 acres, £500,000.

Now the best of the figures can only show one-third of the output of Oathill per acre. Two of these estates are, of course, on some of the finest and richest land in the country. Mr. Paterson would probably have much poor land, and here the pre-war output also quoted, of £28,000 to £30,000 in 1939, shows a turnover of just over fifty *shillings* per acre, against our *pounds* per acre. If anyone should contend that our output is not within the reach of small farmers, I am indebted to Mr. J. F. Cheesewright of Holbeach, Lincs., who has the case of the smallholder very much at heart. In his district there are several thousand acres of smallholdings, many farmed by ex-service men in units of ten acres. After twenty-four years, seventy per cent are still there, only two per cent have been failures. Many own a tractor and a full set of implements, nearly all have a car, and some are on the telephone. He estimates the profit to be at least £15 an acre, and inquires if the 500-acre farmer makes the same. This, of course, is on naturally good land, though in our experience it is possible to do as well on poor land, well farmed and heavily stocked. It is known that the average output per acre was £7 in pre-war days, and it can be calculated from other figures published that this has now risen to £11—that is gross output, not profit—for the whole of England and Wales, so it will be seen there is ample scope for improvement to bring, the general average up to the standard maintained by the smallholders in Lincolnshire, or by stockbreeding on this poor, stony Cotswold land.

As a pupil I saw that profit in farming depended on quick turnover and intensity of production. Our figures over the years have proved it. We have been able to show steadily increasing output, and satisfactory profits even in the depression of the 1930's, so it would appear that our methods have something to recommend them, and if generally adopted might well prove the salvation of English farming and our country. It would also be starting from the right end. There is nothing like successful farming to give a man confidence in himself. Far too often in the last twenty years I have heard some farmer say, 'The Government ought to do something about it'. How often must our friends have heard 'What *we* are going to do this year', possibly *ad nauseam*, but have they ever heard us say 'Let us wait and see what will be arranged at the Ottawa or Sydney Conference'? No! We have been far too busy getting on with the job of breeding better stock or growing finer crops, and most people who study the accounts will agree that our policy was justified.

So deep has the rot taken hold of agriculture that I saw headlines in the *Farmers' Weekly* last year, 'Ought Not Something Be Done About Us?' Reading on I found that the paper had arranged for a group of competition-prize-winning children to visit one of the best Farm Institutes for a short period; and the headline was how one of the children summed up their reaction to the experience, for they had enjoyed themselves and would have liked to take their place in agriculture. The answer is, 'No, my dears, if you cannot find a good old-fashioned farmer to teach you, study the theory and science in your spare time, save your money, and take a farm in due course, then you are not worthy of the finest farming country in the world. A college or farm institute might give you the training which, if you were very clever, might earn a position as a County Organizer, starting at less money than a Corporation bus driver, but only on the land itself can you learn

'That just as the flowers of the garden,
Spring up from the dark, cold earth,
It isn't the smooth, or the easy,
That will make you a man of worth.'

Now turn to the accounts. We have always adopted a fixed method of valuation—and not on the market value of the stock. For the market value system shows a large profit—on paper—in a good time, and a heavy loss—again on paper—in a bad slump; whereas by valuing at a fixed sum for the livestock as shown below, the profit does not appear until the stock is sold, and then it is a cash profit, not a paper one.

## FIXED VALUATIONS

| | |
|---|---|
| Sheep. | Lambs 40/-, 2-tooth ewes, 60/-, 4-tooth 70/-, 6 tooth– 60/-, full mouth 50/-.<br>Rams. Lambs at cost less £2 per annum depreciation. |
| Cattle. | £6 up to six months, £1 per month up to two years.<br>Bulls at cost, less £5 per annum. |
| Poultry. | Hens 5/– each, Pullets 7/6, Stock Cockerels and Cocks the same. Chickens at cost. |
| Pigs. | Sows £10, Boars, at cost less £5 per annum depreciation. Store pigs 6d. per lb. live weight. |

In the case of dead stock we take the previous year's figures (less any sales, plus any purchases) less 10 per cent for depreciation. This does, of course, make a very low figure in our current accounts, for we have many machines which have depreciated 10 per cent per annum for twenty years, but would make more than new price if sold to-day. For example, a machine costing £ 25 when we started, now comes into the valuation at £2 18s. 4d. but would be considered very cheap at £20 now, for it has been kept in good repair and well painted in the original colours. For that reason no-one must think that he could set up the same amount of machinery and poultry appliances for the same figures as we show.

In the following statement of accounts I have taken three typical years: 1924, our first financial year; 1932, the bottom of the agricultural depression; and 1942 for the current position, for it may be assumed that no-one would wish to read through all the figures for twenty years; but I may say they have shown a steadily increasing output and profit over the whole period.

### VALUATIONS

|      | Sheep £ | Cattle £ | Horses £ | Pigs £ | Poultry £ | Dead stock £ |
|------|---------|----------|----------|--------|-----------|--------------|
| 1924 | 56      | 143      | 91       | 13     | 89        | 169          |
| 1932 | 62      | 176      | 28       | 21     | 118       | 247          |
| 1942 | 68      | 457      | 18       | 46     | 293       | 945          |

The 1942 valuation figures are very low indeed as compared with the current market prices, but in the slump of 1932 our fixed prices seemed almost, too high; still, we have kept to them throughout the years and no doubt they will come into line again with market prices in the next slump; then we shall be saved the despair which many farmers experienced of finding they had perhaps worked hard for a year and apparently lost a lot of money as well. What comes in we will have earned, with no extra depreciation to write off.

To turn to the sales, I have again taken the figures for the same three years, but there are two points that need explaining. The sale of horses for £54 in 1924 is accounted for by the fact that we were working the farm with horses and it was possible to make a profit by selling a good one; since then we have changed over to tractor cultivations and keep a cob for taking round poultry food and other light work.

The pig sales in 1942 are very low, but for a short period we were keeping the minimum number in accordance with Government instructions, and to make better use of feeding-stuffs. In 1941 sales amounted to £942, while the figures for 1943 will be well up again as pigs are now wanted, while we have not reared so many pullets owing to the dropping of the Domestic Poultry Replacements Scheme, which left us more food for pigs.

It will also be wondered why corn sales have not improved as much as stock, but in 1932 we consumed the greater part of a good harvest, because the corn prices were too low. In 1942 we sold only wheat, as required by law, keeping back the oats for stock-feeding.

The high miscellaneous sales in 1924 were accounted for by firewood out of the hedges.

### SALES

| | £ Sheep | £ Cattle | £ Horses | £ Pigs | £ Poultry | £ Crops | £ Miscellaneous | £ Total |
|---|---|---|---|---|---|---|---|---|
| 1924 | 90 | 102 | 54 | 76 | 146 | 162 | 75 | 595 |
| 1932 | 125 | 187 | — | 148 | 1,314 | 104 | 27 | 1,905 |
| 1942 | 140 | 1,815 | — | 152 | 5,075 | 174 | 53 | 7,409 |

### EXPENDITURE

| | £ Wages and Insurances | £ Feeding-stuffs | £ Livestock | £ Seeds and Manures | £ Rates, Tithes Interest | £ Miscellaneous | £ Total |
|---|---|---|---|---|---|---|---|
| 1924 | 32 | 63 | 124 | 108 | 115 | 63 | 505 |
| 1932 | 103 | 853 | 163 | 74 | 26 | 88 | 1,307 |
| 1942 | 301 | 1,103 | 721 | 227 | 8½ | 405 | 2,765 |

The experienced farmer will wonder how we managed to spend £1,103 on feeding-stuffs in wartime with the strict rationing in force; the explanation is that we used large quantities of feeding potatoes, weed seeds, and other waste products. In looking down the column I see whey products £35 17s. 4d., which can be taken as an example. This is a waste from cheese-making but a first-class food for pigs and poultry.

The next surprising figure is £721 for livestock in the same year; this is explained by the fact that we spent nearly £500 on day-old chicks. Every one of these chicks was hatched from an egg laid on this farm, but as we had already agreed to supply nearly all our hatching eggs to a well-known hatchery for the year, by the time the Ministry brought out their Domestic Poultry Keepers' Replacement Scheme, we continued to send the eggs to the hatchery and bought the chickens back. We were, of course, paying them a profit for a job which could be carried out here, but as we would have a fair margin on the pullets, they were entitled to their share. In any case we have always honoured our agreements, and consider the satisfaction of. a customer of far greater importance than a profit, an ethical principle of business which has always paid us well. In 1943 the expenditure on chickens was saved as we could give fair notice that we were keeping back a considerable number of eggs for our own machines.

The miscellaneous expenditure also seems to be high, and some may think that they should be entered under different headings, but I give the list in detail as it does illustrate the way in which small expenditures mount up in the course of a year, and these are apt to be over-ooked by those starting farming.

| | £ | s. | d. |
|---|---|---|---|
| Implement repairs | 104 | 1 | 4½ |
| Milk Marketing Board levy | 1 | 0 | 10 |
| Fuel | 83 | 16 | 2½ |
| Carriage | 73 | 2 | 0 |
| Threshing and baling hay | 33 | 9 | 8 |
| Vet. and medicines | 13 | 15 | 6 |
| Licences | 20 | 0 | 0 |
| Registration fees—calves, cattle, pigs | 21 | 3 | 6 |
| Bull licence | 0 | 5 | 0 |
| Rat catching | 5 | 0 | 0 |
| Sack hire | 4 | 6 | 3 |
| Market expenses | 0 | 18 | 9 |
| Accountant's fees | 1 | 0 | 0 |
| Postage | 2 | 0 | 0 |
| Telephone | 4 | 18 | 1 |
| Advertisements | 10 | 17 | 10 |
| Other expenses | 24 | 17 | 4 |
| | £404 | 12 | 4 |

While in 1924 we were paying tithe, rates, and interest on loan, amounting to £115, the equivalent of a high rent for such a derelict place, by 1932 the land had been derated and the loan paid off. I should be mentioned we were responsible for two loans, working capital, and interest on the mortgage with which the farm was bought. By the time we had paid off the first loan we regarded ourselves as established tenant farmers, our landlord being the mortgagee. Then when that was redeemed we became the owner-occupiers. It is often said that a mortgage is a bad landlord. Any sort of landlord was better to us than none at all, and we certainly never kept him waiting a day for his rent, for we did not want to have it called in and incur the heavy charges of finding another mortgage. It is now interesting to recall that our bank manager did not consider us a good enough risk for a mortgage, though the manager at the bank where my father had done business for many years gave us a first-class reference, saying he 'had watched us grow up and we were worthy sons of a mother who had shouldered the responsibilities of an insolvent business, on the death of the father, and every creditor had been paid in full'. I mention this simply to show how one may be helped or handicapped in business, and aspiring young farmers should choose their bank manager as carefully as a wife.

To return to the figures. In due course we redeemed the tithe, leaving only the rates, an item of approximately £8. It is for this reason that we have not included the assessed rental in our figures, as in the early days we were paying a bigger sum in interest than we would have paid as tenants as rent, whereas in later years it should, perhaps, be included, but we look upon it as part of the interest on our capital.

Now we turn to profits, which include our own labour and also interest on the capital sum invested in the holding, amounting to £124 in 1924, £581 in 1932, and £4,484 in 1942. If the reader thinks he can check these figures by deducting the expenditure from the sales I fear that he will be a little out, as I do not give the opening and closing valuations; but the 1942 figures, which to some may appear a misprint, have been checked by the Agricultural Economics Department of Reading University, who show the farm profit for 1942 to be

| | | | |
|---|---|---|---|
| **Farm Profit** | £3,989 | 15 | 0 |
| **Family Labour** | 409 | 0 | 0 |
| **Family income** | £4,398 | 15 | 0 |

To this sum I have added the assessed rental of £85 which they treat as an expenditure, making a total profit, or perhaps, return for capital and labour would be a better word, of £4,484. The figure clearly shows what may be achieved by having a thorough grasp of the underlying principles for successful farming, which I have laid down in this book, and which may be summarized in three words: Work, Muck, and Thought.

# CHAPTER FIFTEEN

## CONCLUSION

And now having told you the history of Oathill, all that remains is to conduct you round the fields and buildings shown in the pictorial map. It was late July when the artist drew the picture and so I describe the farm as it was on that day.

As you came in over the rough common you will have driven over the road we built, probably without a thought, for all farms should have a road to them, but you would have received a very different impression had you ploughed through the swamp in the early-days. The house you pass on the right is the lodge my brother built in the early days of the war.

While you did not think about the road, the white gates swinging easily on their hinges may have caught your eye; these are the results of a queer idea we hold that it is cheaper to repaint a gate every second year than to have it rot or be broken by dragging on the ground in a few months. There are sixteen gates on the farm, and now that good, seasoned wood is so scarce and dear we are glad we have looked after them.

Though in all probability the gates stood open, for the calves you see grazing near the dutch barns are confined with a single, thin strand of wire, electrically charged, and which they graze up to, but do not touch. We are great believers in electric fences; we have six units, and a great length of wire. In the past we often tethered the Jerseys to make the best possible use of the grazing, for it trebles the number which can be carried; but now we only tether bulls, for the rest of the cattle can be folded like sheep with the aid of the electric fencer. For the sheep also the conventional fold of hurdles has disappeared, but these require two strands of wire to keep them in, and they have to learn the lesson that they must not touch the wire before they grow their wool. The saving is considerable for although we can set hurdles at the rate of one a minute, a roll of sheep netting in twenty minutes, six hundred yards of electric fence can be erected in an hour.

On the side of the road you saw potatoes growing, a fair crop for the season, chiefly interesting for the fact that they were planted by Land Girls from the local hostel, achieving in seventeen and a half 'woman' hours what normally requires twenty man hours, and the work was done thoroughly and well, for you see the plants evenly spaced and straight in the rows. I worked with them, always helping the slowest, and cheered them on. The potash manure was very dusty and unpleasant, but I told them it was the basis of a famous beauty preparation ('Not that any of you young ladies would need to use it'), but the next day one of them said that another had washed her face in the water she used for washing her jumper the night before. This shows what a little psychology will do to help on with the good work, for they no longer worried if the manure made their eyes water or faces burn. Normally we do not need extra labour, except for threshing, but as we were held up by rough winds we made use of their labour, being glad of the extra help, for potato-growing is a wartime necessity which interferes with our normal routine.

Above the potatoes you saw the staff hoeing. I mean hoeing, not leaning on their hoes or gazing round the countryside. They probably laughed and chattered as they competed against each other, but the trained worker seldom stops before he reaches the end of the row, and you would see or hear nothing of the three fatal 'S's'. They are hoeing kale, not such a good crop as we would like, but a good wet day would transform it, as it is a full crop, but rather late, for it had to be redrilled on account of the ravages of the turnip fly.

Arriving at the farm, you saw the house standing behind its old spruce firs, which remind me of the mountains, but are probably the relics of some long-past Christmas, for they have grown little in twenty years. The lawn and flower garden are the special care of my sister; while though we manure and dig the vegetable garden my mother plants and weeds it, defying any weed, slug, or insect to appear, and we are seldom short of salads or vegetables, to say nothing of herbs and soft fruits.

Adjoining the house is the orchard, which we replanted in our first year, and although the soil is not well suited to apple-trees, it yields a fair crop in a good season such as this.

Walking through the buildings, for only the youngest calves are in, I trust you see them neat and tidy. They are at least well whitewashed inside, and camouflaged green outside; so well that a pilot once landed his machine in the belief that this was an aerodrome. Unless the trailers and machines are at work they will be under the cartshed or dutch barn, for we hate to see them standing out. I should perhaps mention that, in common with other farms, we have a rubbish dump, but we have planted a screen of poplars and willows round it, so there is no need to fall into the common error of the visitor who congratulates us on the fact that we have not got one. We can even put nettles to good use on this farm so you will not see any clumps in the rickyard, which in fact no longer exists, for we ploughed it in our early days and it has grown an average of twelve loads of mangels on one-fifth of an acre ever since. It took two years to clear, digging up old cart wheels, traction-engine strakes, pitch-forks, parts of machines, and other interesting relics of bygone farming, very dangerous to our horses. We have thought it well worth while, for the tons and tons of old straw, chaff, and weeds which must have rotted down over the ages, make it the richest land on the farm; and this year it promises to grow the proverbial pound of onions to the foot on the six or seven rows we have put in.

And now you pass from the main buildings to the pig-house, which even on a hot July afternoon is quite a pleasant place, for with its insulated walls and roof, and ample ventilation, it is as cool as anywhere on the farm. The brooder-house is empty, the later-hatched chickens being out in the poultry folds which you will see later.

From there we take you to the Spring Field, for here rises the water that supplies the buildings and every field on the farm. It is free from pollution, for we have fenced out the cattle and planted trees round the spring, and it never freezes or runs dry. It is still running several hundred gallons per hour at the end of this long dry spell.

In this field you see nine large poultry-houses which accommodate between them over a thousand hens on free range.

Twenty yearling Jersey heifers gather round you, and it's no use waving your stick at them, for they do not know what a stick is! Any one of them will permit you to place your arm round its neck and read the number tattooed in its ears.

The one you like the look of is 'Enstone Beauty', out of a heifer which gave 800 gallons with her first calf, and has register of merit blood on both sides.

As you walk on they come sedately behind you, for you may be a potential customer, and very early they learn quite a lot about salesmanship. You think they look well on bare pasture? They should; for they are getting green vetches from the arable.

From there you pass through another white gate to Ducks Piece, where there are four milking cows, and fifteen in-calf heifers. Once more they gather round, for what is more annoying than trying to look at a beast that keeps about a hundred yards away all the time? These heifers show more promise for they are autumn calvers—so earnestly required by the Milk Marketing Board—and we trust they will maintain the reputation of Oathill in the herds of the producer-retailers who usually buy from us. We shall want their calves, of course, for our job is to breed and rear good cattle, not to milk them.

By stepping over the electric fence you come to the poultry fold units used for rearing pullets. Each one holds forty to fifty young birds, and is the finest system of rearing we have ever found; they get fresh ground every day and never fail to show their gratitude in the crop that grows after them.

The sheep which have been grazing round the fold units, now gather to meet us, which seems unusual at this time of the year; but in this long drought mangels are being fed again, carefully stored from last winter, and well justifying the trouble in such a season as this, for a sheep never says no to a succulent root when the grass is brown.

Then from here to the next field which is Radford Hill, the largest on the farm. The stubble the tractor is ploughing is vetch, oat, and ryegrass mixture. It gave us grazing last autumn, silage in May, hay in late June, and is now being ripped up in the hope that a thunderstorm will enable us to take a crop of rape before the autumn corn is planted. If not, we shall at least have the benefit of a bastard fallow, for it will be some time before it comes in roots again.

And now across to the oats. Quite useful? Yes. Unless the good thunderstorm we want for the rape comes and knocks them all down, for they are already leaning from their length of straw and weight of grain. Well-manured light soil never seems able to hold up a crop like clay.

And now down to the re-seeded pasture on the steep bank below, where graze the two horses. This too is brown, but at the first wet day will pay tribute to Sir George Stapledon, when we shall see the green shadow creeping over it.

And now back to Garden Ground to see the wheat, which this year is not up to the Oathill standard. It came direct from the National Institute of Agricultural Botany, and was grown under contract. The crop is no reflection on the Institute, but an error of judgement on our part. It had everything a wheat crop should have at planting, and came too well, for we could not make up our minds whether to graze it back with sheep in the spring or not. We left it, the broad flag was attacked by rust, and we have probably lost eight bushels to the acre. On the other hand, had we grazed it and then had a long dry spell, we would have checked it that way. It is one of the most difficult problems to decide on this land, for another farmer who did graze back this year has since told me that he regretted his action. One very old, observant, and experienced farmer once told me that in fifty years of farming he had grazed six times; on three occasions he was glad, and three times he was sorry.

Generally speaking we make one mistake somewhere on our arable each year, however careful we are to try and remedy it in the future.

On the steep ground below the wheat you see more poultry-houses, and these birds will have the run of the stubble and ploughing, until we are ready to plant the winter vetches, for we never leave the land idle a day longer than is necessary.

Then we pass to the workshop, which in normal times was a hive of industry, but the power is now only used for grinding corn and other waste. The building also provides useful storage and any sort of farm or estate repair can be carried out, for it is still fully equipped to make anything in wood or metal. Here also is the office where the records are kept.

And now to the house for a cup of tea, and to meet the staff, for the pupils are the farmers of the future, and therefore the most valuable and important stock on the farm; for it is their youth and energy which have contributed so largely to what you have seen to-day. You have seen nothing very sensational in our methods, just good straightforward mixed farming and useful stock, and a full use made of the land and our opportunities. These are well within the reach of anyone, and if you farm, or intend to farm, your little bit of land as well, or a little better, then our blessings go with you, for we believe you will achieve more for Old England than all the farmers in Bedford Square or Whitehall; for did not the Poet of the People say:

> 'Gi'e fools their gold, and knaves their pow'r,
> Let fortune's bubbles rise and fall.
> Who sows a field or trains a flower
> Or plants a tree, is more than all.'